NOVICE CLASS RADIO AMATEUR FCC TEST MANUAL

by

MARTIN SCHWARTZ

Published by
AMECO PUBLISHING CORP.
220 E. JERICHO TURNPIKE
MINEOLA, NEW YORK 11501

NOVICE CLASS
RADIO AMATEUR
FCC TEST MANUAL

Copyright 1987
by the
Ameco Publishing Corp.

Library of Congress Catalog No. 83–072766

ISBN No. 0–912146–21–4

Printed in the United States of America

PREFACE

This Novice Class FCC Test Manual is part of a series of books published by the Ameco Publishing Corp. for the purpose of preparing individuals for the Federal Communication Commission's Amateur Radio Operator's examinations.

The 300 questions and multiple choice answers in this manual have been issued by the Volunteer Examiner Coordinator Committee under FCC supervision. In those instances where the author feels that the correct multiple choice answer is complete and adequate for the proper understanding of the subject matter, there is no discussion; only the correct answer is indicated. In most of the questions, the discussions explain the correct multiple choice answers and give additional useful material that helps with the understanding of the questions and answers.

In accordance with the rules, the volunteer examiner makes up a 30 question written test, taken from the questions in this manual. There are 30 groups of questions. The examiner selects one question from each group. The format of the questions (multiple choice, essay, or single answer) is at the discretion of the examiner. 74% of the questions must be answered correctly in order to pass.

The numbers at the end of most of the questions refer to the chapters in the Ameco "Novice Class Theory Course" (Cat. #23-01) and the "Ameco Radio Theory Course" (Cat. #102-01), in which more background information on the question can be found. The number to the left of the slash mark indicates the chapter in Cat. #23-01; the number to the right of the slash mark indicates the chapter in Cat. #102-01. These two books are written in an easy-to-understand manner, and assume that the reader has no previous technical knowledge. It is suggested that the reader obtain a copy of these texts so that he may be able to understand basic radio theory sufficiently to enable him to operate his station properly. These books are described on the back cover of this manual.

In order to simplify the work of the examiners, the Ameco Publishing Corp. has published the actual Novice Class operator tests. Each test consists of 30 questions, randomly selected from the questions in this book, and is enclosed in a tamper-proof sealed envelope. The answers are furnished along with each test. A copy of FCC Form 610 is also included. These test kits (Cat. No. EN-1) may be obtained from the Ameco Publishing Corp. or from the various Radio Amateur dealers throughout the United States. See Back Cover for details. GOOD LUCK!

TABLE OF CONTENTS

SUBELEMENT 2A
Rules and Regulations
(9 questions)

One (1) question must be from the following:

Question 2A-1.1. What is the Amateur Radio Service? 14/15*
A. A private radio service used for personal gain and public benefit
B. A public radio service used for public service communications
C. A radio communication service for self-training and technical experimentation
D. A private radio service intended for the furtherance of commercial radio interests
   The answer is C. The Amateur Radio Service is also used for intercommunication between Amateur Radio operators.

Question 2A-2.1. Who is an amateur radio operator? 14/15
A. A person who has not received any training in radio operations
B. Someone who performs communications in the Amateur Radio Service
C. A person who performs private radio communications for hire
D. A trainee in a commercial radio station
   The answer is B. An amateur radio operator is a person holding a valid license to operate an Amateur Radio station issued by the Federal Communications Commission (FCC).

Question 2A-3.1. What is an amateur radio station? 14/15
A. A licensed radio station engaged in broadcasting to the public in a limited and well-defined area
B. A radio station used to further commercial radio interests
C. A private radio service used for personal gain and public service
D. A radio station operated by a person interested in self-training, intercommunication and technical investigation
   The answer is D. An amateur radio station is a station licensed in the amateur radio service embracing necessary apparatus at a particular location used for amateur radio communication.

Question 2A-4.1. What is amateur radiocommunication? 14/15
A. Non-commercial radio communication between Amateur Radio stations with a personal aim and without pecuniary interest
B. Commercial radio communications between radio stations licensed to non-profit organizations and businesses
C. Experimental or educational radio transmissions controlled by student operators
D. Non-commercial radio communications intended for the education

* See Preface

and benefit of the general public

The answer is A. There should be no business interest involved in an amateur radiocommunication.

Question 2A-5.1. What is that portion of an amateur radio license that conveys operator privileges? 14/15
A. The verification section
B. Form 610
C. The operator license
D. The station license
The answer is C.

Question 2A-6.1. What authority is derived from an amateur radio station license? 14/15
A. The authority to use specified operating frequencies
B. The authority to have an Amateur Radio station at a particular location
C. The authority to enforce FCC Rules when violations are noted on the part of other operators
D. The authority to transmit on either amateur or Class D citizen's band frequencies
The answer is B. An amateur radio station license is the instrument of authorization for a radio station in the amateur radio service.

Question 2A-7.1. What is a control operator? 14/15
A. A licensed operator designated to be responsible for the emissions of a particular station
B. A person, either licensed or not, who controls the emissions of an Amateur Radio Station
C. An unlicensed person who is speaking over an Amateur Radio Station's microphone while a licensed person is present
D. A government official who comes to an Amateur Radio Station to take control for test purposes
The answer is A. A control operator is an amateur radio operator designated by the licensee of an amateur radio station to also be responsible for the emissions from that station. The control operator may also be the licensee of the station.

Question 2A-7.2. What is the term for the amateur radio operator designated by the station licensee to also be responsible for the emissions from that station? 14/15
A. Auxiliary operator
B. Operations coordinator
C. Third party
D. Control operator
The answer is D. See answer to question 2A-7.1.

Question 2A-8.1. What is third-party traffic? 14/15
A. A message passed by one Amateur Radio control operator to another Amateur Radio control operator on behalf of another person
B. Public service communications handled on behalf of a minor

political party
C. Only messages that are formally handled through Amateur Radio channels
D. A message from one Amateur Radio station to another in which a third Amateur Radio station must relay all or part of the message because of propagation problems

The answer is A. Third party traffic is amateur radio communication by or under the supervision of the control operator at an amateur radio station on behalf of anyone other than the control operators of the two stations.

Question 2A-8.2. Who is a third-party in amateur radiocommunications? 14/15
A. The Amateur Radio station that breaks into a two-way contact between two other Amateur Radio stations
~ B. Any person passing a message through Amateur Radio communication channels other than the control operators of the two stations handling the message
C. A shortwave listener monitoring a two-way Amateur Radio communication
D. The control operator present when an unlicensed person communicates over an Amateur Radio station

The answer is B. See answer to question 2A-8.1. A third party is the person on whose behalf a message is being sent. A third party may also participate in Amateur Radio. He can talk into the microphone as long as a licensed amateur is operating and controlling the station. In other words, a third party is the person who is NOT OPERATING the station, but who is either participating in the station activities or on whose behalf a message is being sent.

One (1) question must be from the following:

Question 2A-9.1. What are the Novice control operator frequency privileges in the 80 meter band? 14,App.V/15
A. 3500 to 4000 kHz   — B. 3700 to 3750 kHz
C. 7100 to 7150 kHz   D. 7000 to 7300 kHz

The answer is B. He may use radiotelegraphy between 3,700 and 3,750 kHz, using type A1A emission. See the chart on page AP-3.

Question 2A-9.2. What are the Novice control operator frequency privileges in the 40 meter band? 14,App.V/App.8,15.
A. 3500 to 4000 kHz   B. 3700 to 3750 kHz
—C. 7100 to 7150 kHz   D. 7000 to 7300 kHz

The answer is C. He may use radiotelegraphy between 7,100 kHz and 7,150 kHz (7050-7075 kHz when the terrestrial station location is NOT within Region 2), using only type A1A emission. Region 2 includes North and South America.

Question 2A-9.3. What are the Novice control operator frequency

privileges in the 15 meter band? **14,App.V/15,App.8.**
—A. 21.100 to 21.200 MHz   B. 21.000 to 21.450 MHz
C. 28.000 to 29.000 MHz   D. 28.100 to 28.500 MHz
     The answer is A. He may use radiotelegraphy between 21,100 kHz and 21,200 kHz, using only type A1A emission.

Question 2A-9.4. What are the Novice control operator frequency privileges in the 10 meter band? 14,App.V/15,App. 8.
A. 10.100 to 10.109 MHz   B. 10.115 to 10.150 MHz
C. 28.000 to 29.700 MHz   —D. 28.100 to 28.500 MHz
     The answer is D. The Novice control operator frequency privileges in the 10 meter band were increased on March 21, 1987. Prior to that time, he was only permitted to operate CW (A1A) from 28.1 to 28.2 MHz. Now, he can operate CW from 28.1 to 28.5 MHz. He can also use the digital mode (F1B) from 28.1 to 28.3 MHz and sideband telephony (J3E) from 28.3 to 28.5 MHz.

Question 2A-9.5. What, if any, frequency privileges are authorized to Novice control operators beside those in the 80, 40, 15 and 10 meter bands? 14,App.V/15,App.8.
A. All authorized Amateur Radio frequencies above 50.0 MHz
B. None
C. 145 to 147 MHz
—◦ D. 222.1 to 223.91 MHz and 1270 to 1295 MHz
     The answer is D. The Novice enhancement rule of March 21, 1987, gave Novices additional frequency privileges in part of the 1.25 meter band (222.1 to 223.91 MHz) and in part of the 0.23 meter band (1270 to 1295 MHz). Prior to March 21, 1987, Novices were not permitted to operate in these bands.

Question 2A-9.6. In what frequency bands is a Novice authorized to be the control operator of an amateur station? 14, App.V/15,App.8.
A. 1800 to 2000 kHz, 3750 to 3775 kHz, 7100 to 7150 kHz, 21,100 to 21,200 kHz, 28,100 to 28,500 kHz
—B. 3700 to 3750 kHz, 7100 to 7150 kHz, 21,100 to 21,200 kHz, 28.1 to 28.5 MHz, 222.1 to 223.91 MHz, 1270 to 1295 MHz
C. 3.5 to 4.0 MHz, 7.0 to 7.3 MHz, 21.0 to 21.4 MHz, 28.0 to 29.7 MHz, 1240 to 1296 MHz
D. 3.5 to 4.0 MHz, 7.0 to 7.3 MHz, 14.0 to 14.35 MHz, 21.0 to 21.45 MHz, 28.05 to 29.7 MHz, 222.1 to 223.91 MHz
     The answer is B. Privileges on frequencies 28.2 to 28.5 MHz, 222.1 to 223.91 MHz and 1270 to 1295 MHz were added on March 21, 1987. He always had privileges on the other frequencies listed in answer B.

Question 2A-9.7. What does the term <u>frequency</u> band mean?
A. A group of frequencies in which two way contacts are likely to occur during any time of the day
—B. A group of frequencies in which Amateur Radio transmissions are authorized

C. One specific frequency
D. One specific wavelength

The answer is B. A frequency band is a segment or group of adjacent frequencies in the radio frequency spectrum. For instance, the frequencies between 3,500 kHz and 4,000 kHz can be designated as a BAND of frequencies between 3,500 kHz and 4,000 kHz. We have given names to the various ham bands. The names are derived from the wavelengths of the frequencies. For instance, the wavelength of 3,750 kHz is 80 meters. Thus, the amateur band of frequencies between 3,500 kHz and 4,000 kHz is known as the 80 meter band. Note that a Novice class operator has operating privileges only in the 3,700-3,750 kHz portion of the 80 meter band.

The Novice frequencies of 7,100-7,150 kHz are in the 40 meter band, the Novice frequencies of 21,100 kHz to 21,200 kHz are in the 15 meter band, and the Novice frequencies of 28,100 to 28,500 kHz are in the 10 meter band. When we speak of a Novice band or subband, we mean only those frequencies that a Novice class operator can use.

Question 2A-9.8. What does the term frequency privilege mean?
A. The purchase of a frequency for one's use
B. Permission to use a particular frequency
C. A requirement to use a particular frequency
D. Permission to pass routine traffic only on a particular frequency

The answer is B. Frequency privilege means the right to operate in a particular band of frequencies.

Question 2A-9.9. In what meter band is the Novice control operator frequency privilege 3725-kHz?   App.V/15,App.8.
A. 80 meters   B. 40 meters   C. 15 meters   D. 10 meters

The answer is A. The entire 80 meter ham band is from 3,500 to 4,000 kHz. The Novice sub-band is from 3700 to 3750 kHz. See the chart on Page AP-3.

Question 2A-9.10. In what meter band is the Novice control operator frequency privilege 7125-kHz?   App.V/15,App.8.
A. 80 meters   B. 40 meters   C. 15 meters   D. 10 meters

The answer is B. The entire 40 meter ham band is from 7,000 kHz to 7,300 kHz. The Novice sub-band is from 7100 to 7150 kHz. See the chart on page AP-3.

Question 2A-9.11. What frequencies may a Novice control operator use in the amateur 10-meter band?
A. 28.1 to 28.5 MHz   B. 30.1 to 30.5 MHz
C. 27.1 to 27.5 MHz   D. 28.0 to 29.7 MHz

The answer is A. As of March 21, 1987, Novice operators became eligible to use CW (A1A) between 28.1 and 28.5 MHz. They can also use digital (F1B) between 28.1 and 28.3 MHz and single sideband telephony (J3E) between 28.3 and 28.5 MHz. Prior to March 21, 1987, they could only use CW (A1A) between 28.1 and 28.2 MHz.

Question 2A-9.12. What frequencies may a Novice control operator use in the amateur 220-MHz band?
A. 225.0 to 230.5 MHz   B. 222.1 to 223.91 MHz
C. 224.1 to 225.1 MHz   D. 221.2 to 223.0 MHz
    The answer is B. 222.1 to 223.91 MHz is the Novice sub-band in the 1.25 meter band. The entire band is from 220 to 225 MHz. Novices can use all amateur modes and emissions in their sub-band.

Question 2A-9.13. What frequencies may a Novice control operator use in the amateur 1270-MHz band?
A. 1260 to 1270 MHz   B. 1240 to 1300 MHz
C. 1270 to 1295 MHz   D. 1240 to 1246 MHz
    The answer is C. 1270 to 1275 MHz is the Novice sub-band in the 0.23 meter band. The entire band is from 1215 to 1300 MHz. Novices can use all amateur modes and emissions in their sub-band.

Question 2A-9.14. What frequencies may a Novice control operator use in the amateur 23-centimeter band?
A. 1260 to 1270 MHz   B. 1240 to 1300 MHz
C. 1270 to 1295 MHz   D. 1240 to 1246 MHz
    The answer is C. See answer 2A-9.13. 23 centimeters is equal to 0.23 meters which is the 1215 to 1300 MHz band.

One (1) question must be from the following:

Question 2A-10.1.   What emission type is authorized to Novice control operators?   14, App.V/15,App.8.
A. Any emission authorized to the Amateur Radio Service in the 80, 40, 15 and 10 meter CW subbands
B. Any authorized emission used below 29.7 MHz on the Amateur Radio bands
C. All emissions authorized to the Amateur Radio Service on frequencies between 222.1 and 223.91 MHz
D. A3J between 145 and 147 MHz
    The answer is C. Novice class operators are limited to A1A emission in the 80, 40 and 15 meter Novice subbands. They are limited to A1A, F1B and J3E in their 10 meter subband and are permitted all emission modes in their 1.25 and 0.23 meter subbands.

Question 2A-10.2. What does the term A1A emission mean? 9,10/10,11.
A. Extremely strong, copyable signals.
B. A very low Atmospheric Noise Count
C. CW Morse code without audio modulation of the carrier
D. Amplitude modulated radio telephony with only one sideband
    The answer is C. A1A emission is telegraphy. It is the keyed emission of a CW transmitter. There is NO modulation. The transmitter's RF output emission is simply interrupted in a dot-dash sequence.

Question 2A-10.3. What is the symbol for a transmission of telegraphy by on-off keying? 9,10/10,11.
A. A3J   B. F3C   C. J2B   — D. A1A
    The answer is D. See the answer to Question 2A-10.2.

Question 2A-10.4. What does the term CW mean? 9,10/10,11.
A. Calling wavelength   B. Coulombs per watt
C. Continuous wave   D. Continuous wattage
    The answer is C. The term "CW" is the abbreviation for "Continuous Waves". They are waves of constant amplitude. We interrupt these continuous waves with a telegraphy key in a dot-dash sequence to form the letters of the Morse Code.

Question 2A-10.5. What, if any, emission privileges are authorized to Novice control operators beside A1A? 14,App.V/15,App.8.
A. Any emission authorized to the Amateur Radio Service in the 80, 40, 15 and 10 meter CW subbands
B. Any authorized emission used below 29.7 MHz on the Amateur Radio bands
C. All emissions authorized to the Amateur Radio Service on frequencies between 222.1 and 223.91 MHz
D. A3J between 145 and 147 MHz
    The answer is C. Novice control operators are permitted all emission privileges in the 222.1 to 223.91 MHz and the 1270 to 1295 MHz sub-bands. In addition to A1A, they are permitted F1B (digital) between 28.1 and 28.3 MHz. In addition to A1A, they are permitted J3E (sideband telephony) between 28.3 and 28.5 MHz.

Question 2A-10.6. What telegraphy code may a Novice control operator use?
A. Any telegraphy code authorized for use in the amateur bands
B. Only the International Telegraph Alphabet Number Three
C. ASCII, Packet and RTTY
D. Baudot, AMTOR and CW
    The answer is A.

Question 2A-10.7. Which, if any, telegraphy codes may a Novice control operator use beside the international Morse code?
A. Any telegraph code authorized for use in the amateur bands
B. Audio-frequency-shifted CW and AMTOR
C. ASCII, Packet and RTTY
D. Baudot, Amtor and CW
    The answer is A.

Question 2A-10.8. What does the term emission mean? 10/11,15.
A. RF signals transmitted from a radio station
B. Signals refracted by the E layer
C. Filter out the carrier of a received signal
D. Baud rate

The answer is A. Emission means the radio frequency energy coming from the radio station. There are different types of emissions that are permitted to Amateur operators and each type is designated with a symbol. See the chart on Page AP-1 for a list of the various types of emissions that are used.

Question 2A-10.9. What is the term, as used in the Amateur Radio Service Rules, for a transmission from a radio station? 10/11,15.
A. Modulation index   B. Resolution
C. Emission   D. Demodulated envelope
The answer is C. See the answer to Question 2A-10.8.

One (1) question must be from the following:

Question 2A-10.10. What does the term emission privilege mean? 14/15.
A. Permissible class of operator license
B. Permissible type(s) of transmitted signals
C. Permissible frequency of operation
D. Permissible content of communications
The answer is B. Emission privilege means the types of emission permitted to be used by the different classes of amateur operators. For instance, while Novice class operators are only permitted to use type A1A emission in a part of the 40 meter band, General class operators can use additional types, such as A2A or A3E, in other parts of the 40 meter band.

Question 2A-10.11. What emission types are Novice control operators permitted to use on frequencies from 28.3 to 28.5 MHz?
A. All authorized amateur emission privileges
B. A1A and J3E
C. A1A and F1B
D. A1A and F3E
The answer is B. On March 21, 1987, Novices were granted permission to use A1A (CW) and J3E (sideband telephony) on frequencies between 28.3 and 28.5 MHz.

Question 2A-10.12. What emission types are Novice control operators permitted to use on frequencies from 28.1 to 28.3 MHz?
A. All authorized amateur emission privileges
B. F1B and J3E
C. A1A and F1B
D. A1A and J3E
The answer is C. On March 21, 1987, Novices were granted permission to use A1A (CW) and F1B (RTTY) on frequencies between 28.1 and 28.3 MHz. Prior to that time, Novice operators were only permitted to use A1A between 28.1 and 28.2 MHz.

Question 2A-10.13. What emission types are Novice control operators

permitted to use on the amateur 220-MHz band?
A. All amateur emission privileges authorized for use on 220 MHz
B. F1B and J3E
C. A1A and F1B
D. A1A and J3E
The answer is A. Effective March 21, 1987, Novices were permitted all authorized emission modes from 222.1 to 223.91 MHz. This includes CW, AM, FM, Facsimile, RTTY, etc.

Question 2A-10.14. What emission types are Novice control operators permitted to use on frequencies from 1270 to 1295 MHz?
A. All amateur emission privileges authorized for use on 1270 MHz
B. F1B and J3E
C. A1A and F1B
D. A1A and J3E
The answer is A. Effective March 21, 1987, Novices were permitted all authorized emission modes on 1270 to 1295 MHz. This includes CW, AM, FM, Facsimile, RTTY, etc.

Question 2A-10.15. On what frequencies in the 10-meter band are Novice control operators permitted to transmit emission F1B (RTTY)?
A. 28.1 to 28.5 MHz   B. 28.0 to 29.7 MHz
C. 28.1 to 28.2 MHz   D. 28.1 to 28.3 MHz
The answer is D. See answer 2A-10.12.

Question 2A-10.16. On what frequencies in the 10-meter band are Novice control operators permitted to transmit emission J3E (single sideband voice)?
A. 28.3 to 28.5 MHz   B. 28.0 to 29.7 MHz
C. 28.1 to 28.2 MHz   D. 28.1 to 28.5 MHz
The answer is A. See answer 2A-10.11.

Question 2A-10.17. On what frequencies in the 220-MHz band are Novice control operators permitted to transmit emission F3E (FM voice)?
A. 220 to 225 MHz   B. 222.1 to 223.91 MHz
C. 223 to 225 MHz   D. 223.1 to 224.91 MHz
The answer is B. 222.1 to 223.91 MHz is the Novice sub-band part of the 220 to 225 MHz (1.25 meter) band. In this sub-band, Novices are permitted all amateur emission privileges.

Question 2A-10.18. On what frequencies in the 220-MHz band are Novice control operators permitted to transmit emission A1A (CW)?
A. 220 to 225 MHz   B. 222.1 to 223.91 MHz
C. 223 to 225 MHz   D. 223.1 to 224.91 MHz
The answer is B. Prior to March 21, 1987, Novices were not permitted to use this band. See answer 2A-10.17.

Question 2A-10.19. On what frequencies in the 220-MHz band are

Novice control operators permitted to operate packet radio?
A. 220 to 225 MHz   ⁻ B. 222.1 to 223.91 MHz
C. 223 to 225 MHz   D. 223.1 to 224.91 MHz
    The answer is B. Packet radio is a digital mode that is permitted in the entire band. In the Novice sub-band of 222.1 to 223.91 MHz, Novices are permitted all amateur emission privileges.

Question 2A-10.20. On what frequencies in the 1270-MHz band are Novice control operators permitted to transmit emission F3E (FM voice)?
A. 1240 to 1270 MHz   B. 1250 to 1285 MHz
⁻C. 1270 to 1295 MHz   D. 1295 to 1300 MHz
    The answer is C. 1270 to 1295 MHz is the Novice sub-band part of the 0.23 meter band. FM voice is permitted in the entire band and Novices are permitted all amateur emission privileges in their sub-band.

Question 2A-10.21. On what frequencies in the 1270-MHz band are Novice control operators permitted to transmit emission A1A (CW)?
A. 1295 to 1300 MHz   ⁻ B. 1270 to 1295 MHz
C. 1250 to 1285 MHz   D. 1240 to 1270 MHz
    The answer is B. 1270 MHz to 1295 MHz is the Novice sub-band part of the 0.23 meter band. A1A is permitted in the entire band and Novices are permitted all amateur emission privileges in their sub-band.

Question 2A-10.22. On what frequencies in the 1270-MHz band are Novice control operators permitted to operate packet radio?
A. 1295 to 1300 MHz   ⁻ B. 1270 to 1295 MHz
C. 1250 to 1285 MHz   D. 1240 to 1270 MHz
    The answer is B. 1270 to 1295 MHz is the Novice sub-band part of the 0.23 meter band. Packet radio is permitted in the entire band and Novices are permitted all amateur emission privileges in their sub-band.

One (1) question must be from the following:

Question 2A-11.1. Under what circumstances, if any, may the control operator cause unidentified radiocommunications or signals to be transmitted from an amateur station? 14/15.
A. A transmission need not be identified if it is restricted to brief tests not intended for reception by other parties
B. A transmission need not be identified when conducted on a clear frequency or "dead band" where interference will not occur
⁻C. A transmission must be identified under all circumstances
D. A transmission need not be identified unless two-way communications or third-party traffic handling are involved
    The answer is C. Under NO circumstances may a control operator cause unidentified communications to be transmitted.

Question 2A-11.2. What is the meaning of the term <u>unidentified radiocommunications or signals</u>? 14/15.
A. Radiocommunications in which the transmitting station's call sign is transmitted in modes other than CW and voice
B. Radiocommunications approaching a receiving station from an unknown direction
C. Radiocommunications in which the operator fails to transmit his or her name and QTH
D. Radiocommunications in which the transmitting station's call sign is not transmitted
The answer is D. Unidentified radiocommunications or signals are emissions from a station without any station identification from the operator sending them.

Question 2A-11.3. What is the term for transmissions from an amateur station without the required station identification? 14/15.
A. Unidentified transmission B. Reluctance modulation
C. NON emission D. Tactical communication
The answer is A. The term is unidentified radiocommunications or signals.

Question 2A-12.1. Under what circumstances, if any, may the control operator of an amateur station willfully or maliciously interfere with or cause malicious interference to a radiocommunication or signal? 14/15.
A. You may jam another person's transmissions if that person is not operating in a legal manner
B. You may interfere with another station's signals if that station begins transmitting on a frequency already occupied by your station
C. You may never intentionally interfere with another station's transmissions
D. You may expect, and cause, deliberate interference because it is unavoidable during crowded band conditions
The answer is C.

Question 2A-12.2. What is the meaning of the term <u>malicious interference</u>? 14/15.
A. Accidental interference B. Intentional interference
C. Mild interference D. Occasional interference
The answer is B. Malicious interference means to willfully or deliberately interfere with unlawful intention.

Question 2A-12.3. What is the term for transmissions from an amateur station which are intended by the control operator to disrupt other communications in progress? 14/15.
A. Interrupted CW B. Malicious interference
C. Transponded signals D. Unidentified transmissions
The answer is B.

Question 2A-13.1. Under what circumstances, if any, may the control operator cause false or deceptive signals, or communications to be transmitted? 14/15.
- A. Under no circumstances
- B. When operating a beacon transmitter in a "fox hunt" exercise
- C. When playing a harmless "practical joke" without causing interference to other stations that are not involved
- D. When you need to obscure the meaning of transmitted information to ensure secrecy

The answer is A.

Question 2A-13.2. What is the term for a transmission from an amateur station of the word mayday when no actual emergency has occurred? 14/15.
- A. A traditional greeting in May
- B. An Emergency Action System test transmission
- C. False or deceptive signals
- D. "MAYDAY" has no significance in an emergency situation

The answer is C. The transmission of false or deceptive signals is forbidden.

Question 2A-14.1. Under what circumstances, if any, may an amateur station be used to transmit messages for hire? 14/15.
- A. Under no circumstances may an Amateur Radio station be hired to transmit messages
- B. Modest payment from a non-profit charitable organization is permissible
- C. No money may change hands, but a radio amateur may be compensated for services rendered with gifts of equipment or services rendered as a returned favor
- D. All payments received in return for transmitting messages by Amateur Radio must be reported to the IRS

The answer is A. An amateur station shall not be used to transmit or receive messages for hire, nor for communication for material compensation, direct or indirect, paid or promised.

Question 2A-14.2. Under what circumstances, if any, may the control operator be paid to transmit messages from an amateur station? 14/15.
- A. The control operator may be paid if he or she works for a public service agency such as the Red Cross
- B. The control operator may not be paid under any circumstances
- C. The control operator may be paid if he or she reports all income earned from operating an Amateur Radio Station to the IRS as receipt of tax-deductible contributions
- D. The control operator may be paid if he or she works for an Amateur Radio Station that operates primarily to broadcast telegraphy practice and news bulletins for Radio Amateurs

The answer is D. Control operators may normally not be compensated for transmitting messages at an amateur station. However, a club station control operator may be compensated when the club station is operated primarily for the purpose of conducting amateur communication to provide telegraphy practice transmissions intended for persons learning or improving proficiency in the international Morse code, or to disseminate information bulletins consisting solely of subject matter having direct interest to the amateur service.

Control operators may accept compensation only for such periods of time during which the station is transmitting telegraphy practice or bulletins.

One (1) question must be from the following:

Question 2A-15.1. What are the five principles which express the fundamental purpose for which the Amateur Radio Service rules are designed? 14/15.

A. Recognition of emergency communications, advancement of the radio art, improvement of communication and technical skills, increase in the number of trained radio operators and electronics experts, and the enhancement of international good will

B. Recognition of business communications, advancement of the radio art, improvement of communication and business skills, increase in the number of trained radio operators and electronics experts, and the enhancement of international good will

C. Recognition of emergency communications, preservation of the earliest radio techniques, improvement of communication and technical skills, maintain a pool of people familiar with early tube-type equipment, and the enhancement of international good will

D. Recognition of emergency communications, advancement of the radio art, improvement of communication and technical skills, increase in the number of trained radio operators and electronics experts, and the enhancement of a sense of patriotism and nationalism

The answer is A. The five principles are:

(a) Recognition and enhancement of the value of the amateur service to the public as a voluntary, noncommercial, communication service, particularly with respect to providing emergency communications.

(b) Continuation and extension of the amateur's proven ability to contribute to the advancement of the radio art.

(c) Encouragement and improvement of the Amateur Radio Service through rules which provide for advancing skills in both the communication and technical phases of the art.

(d) Expansion of the existing reservoir within the Amateur Radio Service of trained operators, technicians and electronics experts.

(e) Continuation and extension of the amateur's unique ability to enhance international good will.

Question 2A-16.1. Call signs of amateur stations licensed to Novices are from which call sign group?
A. Group A　　B. Group B　　C. Group C　　◦ D. Group D
　The answer is D. The Amateur call sign consists of three parts:
　　(1) the prefix, which can be one or two letters,
　　(2) a single digit (0 through 9), and
　　(3) a suffix of one, two or three letters.
　There are four groups of call signs: Group A, Group B, Group C and Group D. Each group differs from the other in its structure. For instance, Group A has a one or two letter prefix, a digit and a one or two letter suffix. Group B has a two letter prefix, a single digit and a two letter suffix. The Novice class operators receive their calls from Group D. See answer to Question 2A-16.2.

Question 2A-16.2. What is the format of a Group D call sign?
A. Letter-number-letter-letter (ex: K5AA)
B. Letter-letter-number-letter-letter (ex: KA5AA)
C. Letter-letter-number-letter-letter-letter (ex: KA5AAA)
D. Letter-number-letter-letter-letter (ex: K5AAA)
　The answer is C. The Group D format consists of a two letter prefix (starting with K or W), a single digit (0 through 9), and a three letter suffix (AAA through ZZZ). See answer to Question 2A-16.1.

Question 2A-16.3. What are the call sign prefixes for amateur stations licensed by the FCC? 14/15.
A. The letters A, B, C or D only
B. The letters A or U only
C. The letters W or K only
D. The letters A, K, N or W only
　The answer is D. The call sign prefix can be one or two letters. Single letters are either K, N, or W. Two letter combinations are either AA through AL, KA through KZ, NA through NZ, or WA through WZ.

Question 2A-16.4. What determines the number in an amateur station call sign?
A. Call sign district numbers are assigned in such a way as to have approximately equal numbers of radio amateurs in each district
B. Call sign district numbers are assigned in numerical order. When all of the "1-calls" are assigned, the FCC begins issuing "2-calls", and so on
C. Radio amateurs may request specific call sign district numbers for ease in Morse code reception of their calls
D. The station location address given on an applicant's FCC Form 610 determines what call sign district number appears on an applicant's first radio amateur license
　The answer is D. The digit, which is the second part of the call sign, is a single number from 0 through 9. It indicates the geo-

graphical district where the station is located. For instance, in the Continental United States, a 2 indicates New York and New Jersey; a 6 indicates California, a 9 indicates Illinois, Indiana and Wisconsin, etc.

The digits for the various locations are shown in the chart below:

DIGIT	LOCATION
1	Maine, New Hampshire, Vermont, Massachusetts, Rhode Island, Connecticut
2	New York, New Jersey, Guam
3	Pennsylvania, Delaware, Maryland, District of Columbia
4	Virginia, North Carolina, South Carolina, Georgia, Florida, Alabama, Tennessee, Kentucky, Puerto Rico
5	Mississippi, Louisiana, Arkansas, Oklahoma, Texas, New Mexico
6	California, Hawaii
7	Oregon, Washington, Idaho, Montana, Wyoming, Arizona, Nevada, Utah, Alaska
8	Michigan, Ohio, West Virginia
9	Illinois, Indiana, Wisconsin
0	Colorado, Nebraska, North Dakota, South Dakota, Minnesota, Iowa, Missouri, Kansas

Question 2A-17.1. With which amateur stations may an FCC-licensed amateur station communicate? 14/15.
A. All amateur stations
B. All public noncommercial radio stations unless prohibited by the station's government
C. Only with US amateur stations
D. All amateur stations, unless prohibited by the amateur's government

The answer is D. An FCC-licensed amateur radio station may communicate with all other licensed amateur stations. He may NOT communicate with a foreign-licensed amateur if our government or the government of that particular amateur has given notice that it objects to such radiocommunications.

Question 2A-17.2. With which non-amateur radio stations may an FCC-licensed amateur station communicate? 14/15.
A. No non-amateur stations
B. All such stations
C. Only those authorized by the FCC
D. Only those who use the International Morse code

The answer is C. FCC-licensed radio amateurs may communicate with the following groups of non-amateur radio stations in the listed situations:

(1) Stations in other services, licensed by the Commission and

with United States Government stations for civil defense purposes.

(2) Any other station in an emergency.

(3) Other stations, for test purposes, on a temporary basis.

(4) Any station which is authorized by the Commission to communicate with amateur stations.

Question 2A-17.3. Under what circumstances may an FCC-licensed amateur station communicate with another amateur station in a foreign country?   /15.

A. Only when the foreign country uses English as its primary language

B. All the time, except on 28.600 to 29.700 MHz

C. Only when a third party agreement exists between the US and the foreign country

D. At any time unless prohibited by either the US or foreign government

The answer is D. He may communicate with a station in a foreign country, unless the administration of that particular country, or our government, has served notice that it objects to such communications.

Question 2A-17.4. Under what circumstances (other than RACES operation) may an FCC-licensed amateur station communicate with a non-amateur station?   14/15.

A. Anytime

B. Only on permissible frequencies

C. Only on 28.600 to 29.700 MHz

D. Only when the FCC grants authorization for such communications

The answer is D. See the answer to Question 2A-17.2.

Question 2A-17.5. What is the term used in FCC Rules to describe transmitting signals to receiving apparatus while in beacon or radio control operation?   14/15.

A. Multiplex transmissions

B. Duplex transmissions

C. Signal path transmissions

D. One-way transmissions

The answer is D. The term is "one way transmission". Amateur stations may be used for transmitting signals, or communications, or energy, to receiving apparatus for the measurement of emissions, temporary observation of transmission phenomena, radio control of remote objects, and similar experimental purposes, and for the purposes set forth in Paragraph 97.91 of the Rules and Regulations.

One (1) question must be from the following:

Question 2A-18.1. How often must an amateur station be identified? 14/15.

A. At the beginning of the contact and at least every ten minutes during a contact

B. At least once during each transmission
C. At least every ten minutes during a contact and at the end of the contact
D. Every 15 minutes during a contact and at the end of the contact
The answer is C. An amateur station shall be identified by the transmission of its call sign at the end of each communication, and every ten minutes or less during a communication. The word "communication" is understood to mean a completed conversation. During a single communication, two amateurs can switch back and forth many times.

Question 2A-18.2. If you were an amateur operator, how would you correctly identify your amateur station communications?
A. With the name and location of the control operator
- B. With the call sign of the station licensee in all cases
C. With the call of the control operator, even when he/she is visiting another radio amateur's station
D. With the name and location of the station licensee, followed by the two-letter designation of the nearest FCC Field Office
The answer is B. Identification is made in accordance with the answer to Question 2A-18.1. Your call sign should be given in plain English or international Morse code.

Question 2A-18.3. What station identification, if any, is required at the beginning of a QSO? 14/15.
A. The operator originating the contact must transmit both call signs
- B. No identification is required at the beginning of the contact
C. Both operators must transmit their own call signs
D. Both operators must transmit both call signs
The answer is B. Station identification is NOT required at the beginning of a QSO. See question 2A-18.1.

Question 2A-18.4. What station identification, if any, is required at the end of a QSO?
- A. Both operators must transmit their own call sign
B. No identification is required at the end of the contact
C. The operator originating the contact must always transmit both call signs
D. Both operators must transmit their own call sign followed by a two-letter designator for the nearest FCC field office
The answer is A. Station identification IS required at the end of a QSO. See question 2A-18.1.

Question 2A-18.5. What do the FCC rules for amateur station identification require? 14/15.
A. Each Amateur Radio station shall give its call sign at the beginning of each communication, and every ten minutes or less during a communication
- B. Each Amateur Radio station shall give its call sign at the end of

each communication, and every ten minutes or less during a communication

C. Each Amateur Radio station shall give its call sign at the beginning of each communication, and every five minutes or less during a communication

D. Each Amateur Radio station shall give its call sign at the end of each communication, and every five minutes or less during a communication

The answer is B. See question 2A-18.1.

Question 2A-18.6. What is the fewest number of times an amateur station must transmit its station identification during a 15 minute QSO? 14/15.

A. 1   B. 2   C. 3   D. 3

The answer is B. He must identify first after 10 minutes, and a second time at the end of the QSO.

Question 2A-18.7. What is the fewest number of times an amateur station must transmit its station identification during a 25 minute QSO? 14/10.

A. 1   B. 2   C. 3   D. 4

The answer is C. He must identify first after 10 minutes, a second time after 20 minutes, and a third time at the end of the QSO.

Question 2A-18.8. What is the fewest number of times an amateur station must transmit its station identification during a 35 minute QSO? 14/10.

A. 1   B. 2   C. 3   D. 4

The answer is D. He must identify first after 10 minutes, a second time after 20 minutes, a third time after 30 minutes, and a fourth time at the end of the QSO.

Question 2A-18.9. What is the longest period of time during a QSO that an amateur station does not need to transmit its station identification? 14/10.

A. 5 minutes   B. 10 minutes   C. 15 minutes   D. 20 minutes

The answer is B. See answer to question 2A-18.1.

Question 2A-18.10. What is the fewest number of times an amateur station must identify itself during a 5 minute QSO? 14/10.

A. 1   B. 2   C. 3   D. 4

The answer is A. He must identify once, at the end of the QSO.

One (1) question must be from the following:

Question 2A-19.1. What amount of transmitting power may an amateur station use? 14/15.

A. 200 watts input

B. 200 watts output

C. 1500 watts PEP output

⟶ D. The minimum legal power necessary to maintain reliable communications

The answer is D. Operating with more power than is necessary to maintain reliable communications will cause unnecessary interference to other communications.

Question 2A-19.2. What is the maximum transmitting power ever permitted to be used at an amateur station transmitting on frequencies available to Novice control operators? 14/15.

A. 75 watts PEP output on the 80, 40 and 15-meter bands

B. 100 watts PEP output on the 80, 40 and 15-meter bands

C. 200 watts PEP output on the 80, 40 and 15-meter bands

D. 1500 watts PEP output on the 80, 40 and 15-meter bands

The answer is C. The maximum transmitting power (peak envelope power output) permitted to all amateurs in the 80, 40 and 15 meter Novice sub-bands is 200 watts. Novice and Technician operators are permitted a maximum of 200 watts PEP output in the 10 meter Novice sub-band. Other amateur operators are permitted full power in the 10 meter Novice sub-band. Novice operators are permitted a maximum of 25 watts in the 1.25 meter Novice sub-band and 5 watts in the 0.23 meter Novice sub-band. All other amateur operators may use full power in the 1.25 meter and 0.23 meter Novice sub-bands.

Question 2A-19.3. What is the amount of transmitting power that an amateur station must never exceed when transmitting on 3725-kHz?

A. 75 watts PEP output   B. 100 watts PEP output

C. 200 watts PEP output   D. 1500 watts PEP output

The answer is C. 3725 kHz is a frequency in the 80 Meter Novice sub-band. Therefore, the transmitting power (peak envelope power output) must never exceed 200 watts. See answer 2A-19.2.

Question 2A-19.4. What is the amount of transmitting power that an amateur station must never exceed when transmitting on 7125-kHz?

A. 75 watts PEP output   B. 100 watts PEP output

C. 200 watts PEP output   D. 1500 watts PEP output

The answer is C. 7125 kHz is a frequency in the 40 Meter Novice sub-band. Therefore, the transmitting power (peak envelope power output) must never exceed 200 watts. See answer 2A-19.2.

Question 2A-19.5. What is the maximum transmitting power permitted an amateur station with a Novice control operator transmitting on the amateur 10-meter band?

A. 25 watts PEP output   B. 200 watts PEP output

C. 1000 watts PEP output   C. 1500 watts PEP output

The answer is B. The rule change of March 21, 1987 that expanded the 10 meter Novice sub-band from 28.1-28.2 MHz to 28.1-28.5 MHz did not change the amount of power that a Novice could use in this sub-band. It was and is 200 watts PEP.

Question 2A-19.6. What is the maximum transmitting power permitted an amateur station with a Novice control operator transmitting on the amateur 220-MHz band?
A. 5 watts PEP output   B. 10 watts PEP output
C. 25 watts PEP output   D. 200 watts PEP output
    The answer is C. Amateur operators, other than Novices, are permitted 1500 watts PEP output in the 220-MHz band. However, Novices are only permitted 25 watts PEP output in their sub-band.

Question 2A-19.7. What is the maximum transmitting power permitted an amateur station with a Novice control operator transmitting on the amateur 1270-MHz band?
A. 5 milliwatts PEP output   B. 500 milliwatts PEP output
C. 1 watt PEP output   D. 5 watts PEP output
    The answer is D. Amateur operators, other than Novices, are permitted 1500 watts PEP output in the 1270-MHz band. However, Novices are only permitted 5 watts PEP output in their sub-band.

Question 2A-19.8. What amount of transmitting power may an amateur station with a Novice control operator use on the amateur 220-MHz band?
A. Not less than 5 watts PEP output
B. The minimum legal power necessary to maintain reliable communications
C. Not more than 50 watts PEP output
D. Not more than 200 watts PEP output
    The answer is B. Using the minimum legal power necessary to maintain reliable communications is the rule on all bands and with all classes of Amateur operators.

Question 2A-20.1. If you were an amateur operator and you received an Official Notice of Violation from the FCC, how promptly must you respond? 14/15.
A. Within 90 days   B. Within 30 days
C. Within 10 days   D. The next day
    The answer is C.

Question 2A-20.2. If you were an amateur operator and you received an Official Notice of Violation from the FCC, to whom must you respond? 14/15.
A. Any office of the FCC
B. The Gettysburg, PA office of the FCC
C. The Washington, DC office of the FCC
D. The FCC office that originated the notice
    The answer is D. You must send a written answer directly to the office of the Commission originating the official notice.

Question 2A-20.3. If you were an amateur operator and you received an Official Notice of Violation from the FCC relating to a violation

that may be due to the physical or electrical characteristic of your transmitting apparatus, what information must be included in your response? 14/15.
A. The make and model of the apparatus
B. The steps taken to guarantee future violations
C. The date that the apparatus was returned to the manufacturer
D. The steps taken to prevent future violations
    The answer is D. If the notice relates to some violation that may be due to the physical or electrical characteristics of the transmitting apparatus, the answer shall state fully what steps, if any, were taken to prevent future violations, and if any new apparatus is to be installed, the date such apparatus was ordered, the name of the manufacturer, and promised date of delivery.
    If the notice of violation relates to some lack of attention to, or improper operation of the transmitter, the name of the operator in charge shall be given.

Question 2A-21.1. Who is held responsible for the proper operation of an amateur station? 14/15
A. The control operator
B. The licensee
C. Both the control operator and the licensee
D. The person who owns the property where the station is located
    The answer is C.

Question 2A-21.2. When must an amateur station have a control operator? 14/15
A. A control operator is only required for training purposes
B. Whenever the station receiver is operated
C. Whenever the transmitter is operated, except when the station is under automatic control
D. A control operator is not required
    The answer is C. Every amateur radio station, when in operation, must have a control operator at an authorized control point. The control operator must be on duty, except when the station is operated under automatic control.

Question 2A-21.3. Who may be the control operator of an amateur station? 14/15.
A. Any person over 21 years of age
B. Any licensed Amateur Radio operator
C. Any licensed Amateur Radio operator with an Advanced class license or higher
D. Any person over 21 years of age with a General class license or higher
    The answer is B. The control operator may be the station licensee, if he is a licensed Amateur radio operator, or he may be another Amateur radio operator with the required class of license and designated by the station licensee.

One (1) question must be from the following:

Question 2A-22.1. What does the term "digital communications" refer to?
A. Amateur communications that are designed to be received and printed automatically
B. Amateur communications sent in binary-coded decimal format
C. A "hands-on" communications system requiring manual control
D. A computer-controlled communications system, requiring no operator control

The answer is A. Radioteletype (RTTY) is one of several forms of digital communication. In RTTY, a keyboard at the transmitter, converts letters of the alphabet and numbers into a simple digital code made up of various combinations of "power on-power off" bits. This signal is transmitted and is received by a receiver which decodes the digital signal and converts it back into its original letters and numbers and automatically prints them out.

Question 2A-22.2. What term is used to describe amateur communications intended to be received and printed automatically?
A. Teleport communications   B. Direct communications
C. Digital communications   D. Third-party communications

The answer is C. See answer 2A-22.1

Question 2A-22.3. What term is used to describe amateur communications for the direct transfer of information between computers?
A. Teleport communications   B. Direct communications
C. Digital communications   D. Third-party communications

The answer is C. Amateur radio operators use computers to encode and decode digital signals and communicate with each other in this manner.

Question 2A-23.1. When must the licensee of an Amateur Radio station in portable or mobile operation notify the FCC of such operation?
A. 1 week in advance, if the operation will last for more than 24 hours
B. FCC notification is not required for portable or mobile operation
C. 1 week in advance, if the operation will last for more than a week
D. 1 month in advance of any portable or mobile operation

The answer is B. There are, however, certain conditions which must be met when operating portable or mobile outside the jurisdiction of the United States.

Question 2A-23.2. When may you operate your Amateur Radio station at a location other than the one listed on your station license?
A. Only during times of emergency
B. Only after giving proper notice to the FCC

C. During an emergency or an FCC approved emergency preparedness drill

D. Whenever you want to

The answer is D. See answer 2A-23.1.

SUBELEMENT 2B
Operating Procedures
(2 questions)

One (1) question must be from the following:

Question 2B-1.1. What does the S in the RST signal report mean? 13/14.
A. The scintillation of a signal
⊳B. The strength of the signal
C. The signal quality
D. The speed of the CW transmission
    The answer is B. The RST reporting system is a means of rating the quality of a signal on a numerical basis.
    The S stands for signal strength, and is rated on a scale of 1 to 9. The higher the number, the greater is the signal strength.
    The complete RST reporting system is given in Appendix 2, page AP-2.

Question 2B-1.2. What does the R in the RST signal report mean? 13/14.
A. The recovery of the signal
B. The resonance of the CW tone
C. The rate of signal flutter
⊾ D. The readability of the signal
    The answer is D. The R stands for readability and is rated on a scale of 1 to 5. The higher the number, the better is the readability. See Appendix 2.

Question 2B-1.3. What does the T in the RST signal report mean? 13/14.
⌣ A. The tone of the signal
B. The closeness of the signal to "telephone" quality
C. The timing of the signal dot to dash ratio
D. The tempo of the signal
    The answer is A. The T indicates the quality of a CW tone and is rated on a scale of 1 to 9. The higher the number, the cleaner is the tone. The "T" report is used only for CW signals.
    If the tone is really clean, as though it came from a crystal controlled transmitter, an "X" may be added to the "T" report.

Question 2B-2.1. At what telegraphy speed should a CQ message be transmitted? 13/14.
A. Only speeds below five wpm
B. The highest speed your keyer will operate
⌐ C. The speed at which you can reliably receive
D. The highest speed at which you can control the keyer

The answer is C. An operator should send a CQ at the speed that he can receive CW. The other station will generally answer a CQ at the speed of the CQ. If an amateur sends CQ at a speed higher than he can copy, he will have difficulty when the other station comes back to him at the speed of the CQ.

In general, an amateur should always lower his code speed to accommodate the other amateur.

Question 2B-3.1. What is the meaning of the term zero beat? 12/12.
A. Transmission and reception on the same operating frequency
B. Transmission on a predetermined frequency
C. Used only for satellite reception
D. Unimportant for CW operations

The answer is A. When two signals are mixed or "beat" together, two new signals are produced. The frequency of one of the new signals is equal to the sum of the frequencies of the original signals. The frequency of the other new signal is equal to the difference of the frequencies of the original signals. It is this latter signal that we are interested in. If the difference of the two frequencies is in the audible range, we can hear it. As we move one of the original signals closer in frequency to the other signal, the audible tone becomes lower and lower in frequency. When the frequencies of the two original signals are exactly equal, the difference is zero and we hear nothing. This point is referred to as **"zero-beat".**

The principle of "zero-beating" is used in frequency measuring equipment when we want to find the frequency of an unknown signal. We take a signal with a known frequency and adjust it gradually until it zero-beats with the unknown frequency. At the zero-beat point, we look at the frequency of the known signal (it is usually shown on a calibrated dial or scale). This is also the frequency of the unknown signal.

Question 2B-3.2. Why should amateur radio stations in communication with each other zero beat?
A. Reduction of interference caused by heterodyning carriers
B. Conservation of radio frequency power output
C. Facilitation of synchronous demodulation of A1A emissions
D. Conservation of radio spectrum space

The answer is D. They do this to determine that they are on the same frequency. When an operator "zero-beats" his signal with another station, he knows that he is on the same frequency as the other station. Also, the two stations now occupy a single channel of frequency space in the band, thereby taking up less frequency space than if they were occupying different frequencies. This makes for less interference and allows more radio stations to use the band.

When an operator hears CQ, he quickly zero-beats his transmitter with that of the CQ sender. He does this because he knows that the person sending the CQ will first listen on his own frequency. He, therefore, has a better chance of making the contact.

Question 2B-4.1. How can on-the-air transmitter tune-up be kept as short as possible? 11/12.
A. By using a random wire antenna
B. By tuning up on 40 meters first, then switching to the desired band
C. By tuning the transmitter into a dummy load
D. By using twin lead instead of coaxial-cable feed lines

The answer is C. A dummy antenna should be used in place of the actual antenna during testing and tune-up procedures. A dummy antenna is a resistance load that presents the same resistance and power dissipation to the final stage as the antenna does. A dummy antenna is located right in the amateur station's operating room and converts the transmitter's output power into heat. This prevents the signals from getting out on the air during these procedures. After the tuning and testing procedures have been completed using the dummy load, the antenna can be switched in and the final "touch-up" tuning, which should take very little time, can then be done.

It is a good idea to reduce power to the final stage when tuning up. This will prevent possible damage to the equipment.

Question 2B-5.1. What is the difference between the telegraphy abbreviations CQ and QRZ? 13/14.
A. CQ means "end of contact"; QRZ means "my time zone is ... "
B. CQ means "calling any station"; QRZ means "is this frequency in use?"
C. CQ means "calling any station"; QRZ means "who is calling me?"
D. CQ means "call on each quarter hour"; QRZ means "my radio zone is ..."

The answer is C. CQ is a general inquiry call that is used when an amateur wants to make a contact.

QRZ? means "Who is calling me?" This is usually done when an amateur thinks he hears someone calling him and has missed the call sign of the other amateur.

Question 2B-5.2. What is the difference between the telegraphy abbreviations K and SK? 13/14.
A. K means "all received correctly"; SK means "received some correctly"
B. K means "any station transmit"; SK means "end of contact"
C. K means "end of message"; SK means "best regards"
D. K means "specific station transmit"; SK means "wait"

The answer is B. K means "go ahead". It is used after a CQ, or at the end of a single transmission. SK indicates the end of the **entire** transmission or contact.

Question 2B-5.3. What are the meanings of telegraphy abbreviations DE, AR, and QRS? 13/14.

A. DE means "received all correctly"; AR means "only the called station transmit"; QRS means "interference from static"

B. DE means "calling any station"; AR means "received all correctly"; QRS means "send RST report"

C. DE means "from", or "this is"; AR means "end of message"; QRS means "send more slowly"

D. DE means "directional emissions"; AR means "best regards"; QRS means "radio station location is ..."

The answer is C. DE means "from". It is followed by the call letters of the station doing the sending.

AR indicates the end of a message unit or transmission.

Abbreviations starting with a Q are known as Q signals. When a Q signal is followed by a question mark, it takes the form of a question. When the Q signal is not followed by a question mark, it indicates either a reply to a Q signal question or a direct statement. For instance, QRS? means "Shall I send slower?" QRS without the question mark means "Send slower".

Question 2B-6.1. What is the format of a standard radiotelephone CQ call?

A. Transmit the phrase "CQ" three times, followed by "this is", followed by your call sign three times

B. Transmit the phrase "CQ" at least ten times, followed by "this is", followed by your call sign two times

C. Transmit the phrase "CQ" at least five times, followed by "this is", followed by your call sign once

D. Transmit the phrase "CQ" at least ten times, followed by "this is", followed by your call sign once

The answer is A. Sending CQ more than three times wastes time and is an unnecessary use of the airways. If it is sent less than three times, it will easily be missed.

Question 2B-7.1. How is the call sign "KA3BGQ" stated in Standard International Phonetics?

A. King America Three Baker Golf Queen

B. Kilo Alfa Three Bravo Golf Quebec

C. Kilowatt Alfa Three Bravo George Queen

D. Kilo America Three Baker Golf Quebec

The answer is B. If conditions are poor during telephony operation, or if you feel that you may not be understood, phonetics should be used. The International Telecommunication Union (ITU) recommends the following phonetic alphabet:

A - Alpha	G - Golf	N - November	U - Uniform
B - Bravo	H - Hotel	O - Oscar	V - Victor
C - Charlie	I - India	P - Papa	W - Whiskey
D - Delta	J - Juliette	Q - Quebec	X - Xray
E - Echo	K - Kilo	R - Romeo	Y - Yankee
F - Foxtrot	L - Lima	S - Sierra	Z - Zulu
	M - Mike	T - Tango	

Question 2B-7.2. How is the call sign "WB2OSQ" stated in Standard International Phonetics?
A. Whiskey Baker Two Oscar Sierra Queen
B. Whiskey Bravo Two Oscar Sierra Quebec
C. Willie Baker Two Ontario Sugar Quebec
D. Washington Bravo Two Oscar Sugar Queen
    The answer is B. See answer 2B-7.1.

Question 2B-7.3. How is the call sign "ON4UN" stated in standard international phonetics?
A. Ontario Nancy Four Uncle Nancy
B. Ocean Norway Four Uniform Norway
C. Oscar November Four Uniform November
D. Oscar Nancy Four Unicorn Nancy
    The answer is C. See answer 2B-7.1.

Question 2B-7.4. How is the call sign "WB1EYI" stated in Standard International Phonetics?
A. Whiskey Bravo One Echo Yankee India
B. Whiskey Baker One Echo Yankee Ida
C. Willie Baker One Echo Yankee India
D. Washington Baltimore One Easy Yellow Ida
    The answer is A. See answer 2B-7.1.

One (1) question must be from the following:

Question 2B-8.1. What is the format of a standard RTTY CQ call?
A. Transmit the phrase "CQ" at least ten times, followed by "this is", followed by your call sign two times
B. Transmit the phrase "CQ" at least five times, followed by "this is", followed by your call sign once
C. Transmit the phrase "CQ" three to six times, followed by "DE", followed by your call sign three times
D. Transmit the phrase "CQ" at least ten times, followed by "this is", followed by your call sign once
    The answer is C.

Question 2B-8.2. What are three common sending speeds for RTTY signals on the 10-meter band?
A. "45 speed" (45 bauds), "100 speed" (100 bauds) and "1200 speed" (1200 bauds)
B. "75 speed" (45 bauds), "110 speed" (80 bauds) and "1200 speed" (1170 bauds)
C. "60 speed" (45 bauds), "105 speed" (80 bauds) and "1500 speed" (1475 bauds)
D. "60 speed" (45 bauds), "75 speed" (56 bauds) and "100 speed" (75 bauds)
    The answer is D. The number in front of "speed" indicates the words per minute. For instance, "60 speed" means 60 words per

minute. Baud is a unit of signaling rate.

Question 2B-8.3. What is the commonly used RTTY sending speed above 50 MHz?
A. 1200 bauds   B. 60 bauds
D. 100 bauds   D. 9600 bauds
    The answer is A. The higher the frequency, the higher is the permitted sending speed.

Question 2B-8.4. What is one common use for a RTTY mailbox?
A. To leave a message with an amateur equipment dealer, ordering a new radio
B. Storing messages from one amateur for later retrieval by another amateur
C. To establish a QSO with another amateur RTTY station, and then to move off frequency
D. To leave messages that will be mailed to another person the next day
    The answer is B. An RTTY mailbox is an electronic "mailbox". Messages can be entered into the computer of the mailbox using a specific code. Messages can later be retrieved from the mailbox with the aid of the correct code or password.

Question 2B-8.5. What is the term used to describe an automatic RTTY system used to store messages from amateurs for later retrieval by other amateurs?
A. A message delivery system
B. An automatic teleprinting system
C. A digipeater
D. A RTTY mailbox
    The answer is D. See answer 2B-8.4.

Question 2B-9.1. What do the letters "TNC" stand for?
A. Terminal-Node Controller   B. Tucson Network Controller
C. Terminal Network Contact   D. Tactical-Number Controller
    The answer is A. The TNC is used in Packet radio. It is the heart of the system. It is the section between the computer terminal and the transceiver. It is in the TNC where the information to be transmitted is assembled, packaged and addressed. The "packets" are then passed on to be transmitted. The TNC also receives packets and extracts the information from them.

Question 2B-9.2. What does the term "connected" mean in a packet-radio link?
A. A telephone link has been established between two amateur stations
B. An Amateur Radio message has reached a station for local delivery
C. The transmitting station is sending data specifically addressed to the receiving station, and the receiving station is acknowledging

that the data has been received correctly
D. A transmitting and a receiving station are using a certain digipeater, so no other contacts can take place until they are finished
   The answer is C. "Connected" indicates that a contact has been made and messages can be exchanged. Packet radio uses an automatic system where the stations are continuously checking each other to make certain that everything is being received O.K.

Question 2B-9.3. What does the term "monitoring" mean on a frequency used for packet radio?
A. The FCC is copying all messages, to determine their content
B. A member of the Amateur Auxiliary to the FCC's Field Operations Bureau is copying all messages to determine their content
C. The receiving station's video monitor is displaying all messages intended for that station
D. The receiving station is displaying information that may not be addressed to that station, and is not acknowledging correct receipt of the data
   The answer is D.

Question 2B-9.4. What is a digipeater?
A. A packet-radio station used to retransmit data specifically addressed to be retransmitted by that station
B. An Amateur Radio repeater designed to retransmit all audio signals in a digital form
C. An Amateur Radio repeater designed using only digital electronics components
D. A packet-radio station that retransmits any signals it receives
   The answer is A. A digipeater is actually a digital repeater. It retransmits information received. It does not require an operator to be present; however, its power must be on. The "digipeat" command can be turned off if the operator doesn't wish to use the station as a digipeater.

Question 2B-9.5. What is the meaning of the term network in packet radio?
A. A system of telephone lines interconnecting packet-radio stations to transfer data
B. A method of interconnecting packet-radio stations so that data can be transferred over long distances
C. The interlaced wiring on a terminal-node controller board
D. The terminal-node controller function that automatically rejects another caller when the station is connected
   The answer is B. The network involves a large number of stations using common protocols.

Question 2B-9.6. What is the term used to describe a packet-radio station used to retransmit data specifically addressed to be retrans-

mitted by that station?
A. A RTTY mailbox
B. A network-node controller
C. An autopatch
D. A digipeater
The answer is D. See answer 2B-9.4.

Question 2B-9.7. What is the term used to describe a method of interconnecting packet-radio stations so that data can be transferred over long distances?
A. Networking
B. Crosslinking
C. Autopatching
D. Duplexing
The answer is A. See question 2B-9.5.

Question 2B-9.8. What sending speed is commonly used for packet-radio transmissions on the 220-MHz band?
A. 45 bauds   B. 110 bauds   C. 1200 bauds   D. 12,000 bauds
The answer is C.

Question 2B-10.1. What is a good way to establish a contact on a repeater?
A. Give the call sign of the station you want to contact 3 times
B. Call the other operator by name, then give your call sign 3 times
C. Say, "Breaker breaker", and then give your call sign
D. Call the desired station and then identify your own station
The answer is D. A repeater is a receiver-transmitter device that receives a signal on one frequency and automatically retransmits it on another frequency. The repeater is located on a hill or other high point, and its purpose is to extend the range of communications of low power hand-held and mobile stations. The low power station transmits to a nearby repeater. The repeater then retransmits the signal over a much greater distance than the low power station could transmit.

If you wish to use a repeater, you should, with a brief call, make your presence known. Do not call CQ. If you wish to call someone, then call that station and give your own call.

Question 2B-10.2. What is the main purpose of a repeater?
A. Repeaters extend the operating range of portable and mobile stations
B. To provide a station that makes local information available 24 hours a day
C. To provide a means of linking Amateur Radio stations with the telephone system
D. To retransmit NOAA weather information during severe storm warnings
The answer is A. See answer 2B-10.1.

Question 2B-10.3. Why is there an <u>input</u> and an <u>output</u> frequency to describe the operating frequency of any repeater?
A. All repeaters offer a choice of two operating frequencies, in case

one is busy
B. The repeater receives on one frequency and transmits on another
C. One frequency is used to control repeater functions and the other frequency is the one used to retransmit received signals
D. Repeaters require an access code to be transmitted on one frequency while your voice is transmitted on the other
The answer is B. See answer 2B-10.1.

Question 2B-10.4. When should simplex operation be used instead of a repeater?
A. Whenever greater communications reliability is needed
B. Whenever you need someone to make an emergency telephone call
C. Whenever a contact is possible without using a repeater
D. Whenever you are traveling and need some local information
The answer is C. Simplex means "alternating transmissions between two or more stations using one frequency". It is communicating with another station directly instead of using a repeater. Simplex should always be used, where possible, instead of tying up a repeater.

Question 2B-10.5. What is an autopatch?
A. A repeater feature that automatically selects the strongest received signal to be repeated
B. An automatic system of connecting a mobile station to the next repeater as it moves out of range of the first
C. A system that automatically locks other stations out of the repeater when there is a QSO in progress
D. A device that allows repeater users to make telephone calls from their portable or mobile stations
The answer is D. An autopatch is a device that interconnects between the repeater station and the telephone lines. Thus, an amateur who is in contact with a repeater from his portable or mobile station, can contact someone by telephone via the repeater.

Question 2B-10.6. What is the term used to describe a device that allows repeater users to make telephone calls from their portable or mobile stations?
A. An amateur phone controller B. An autopatch
C. A terminal node controller   D. A phone patch
The answer is B. See answer 2B-10.5.

One (1) question must be from the following:

Question 2C-1.1. What type of propagation uses radio signals refracted back to earth by the ionosphere? 11/12.
A. Skip   B. Earth-moon-earth
C. Ground wave   D. Tropospheric
    The answer is A. The term that is given to this type of propagation is **Ionospheric Propagation.** Ionospheric propagation is the propagation of a sky wave that leaves the antenna and travels skyward until it reaches the ionosphere. Here, it is reflected (technically, the term should be **"refracted"**) back to earth some distance away. See figure 2C-1.1.
    The terms **Sky Wave Propagation** and **Skip Propagation** are also used to indicate ionospheric propagation.

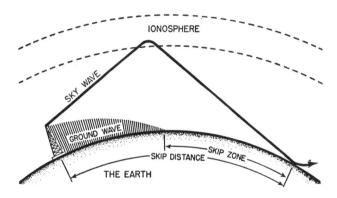

Fig. 2C-1.1. Propagation of Radio Waves.

Question 2C-1.2. What is the meaning of the term skip propagation?
A. Signals reflected from the moon
B. Signals refracted by the ionosphere
C. Signals refracted by water-dense cloud formations
D. Signals retransmitted by a repeater
    The answer B. Skip propagation is another term for ionospheric propagation. Sky wave propagation is a term also used for ionospheric propagation.

Question 2C-1.3. What is the area of weak signals between the ranges of ground waves and the first-hop called? 11/12.
A. The skip zone   B. The hysteresis zone
C. The monitor zone   D. The transequatorial zone

The answer is A. See Figure 2C-1.1.

Question 2C-1.4. What is the meaning of the term skip zone? 11/12.
A. An area covered by skip propagation
B. The area where a satellite comes close to the earth, and skips off the ionosphere
C. An area that is too far for ground-wave propagation, but too close for skip propagation
D. The area in the atmosphere that causes skip propagation
The answer is C. The **Skip Zone** is the area between the end of the ground wave zone and the point where the sky wave returns to earth. See Figure 2C-1.1.

Question 2C-1.5. What does the term skip mean? 11/12.
A. Signals are reflected from the moon
B. Signals are refracted by water-dense cloud formations
C. Signals are retransmitted by repeaters
D. Signals are refracted by the ionosphere
The answer is D. **Skip** refers to the fact that the sky wave travels up to the ionosphere and is refracted back to earth, some distance away from the transmitter. It thus "skips" over a portion of the earth. The transmitted signal cannot be heard in the skip zone. See Figure 2C-1.1.

Question 2C-1.6. What type of radio wave propagation makes it possible for amateur stations to communicate long distances? 11/12.
A. Direct-inductive propagation   B. Knife-edge diffraction
C. Ground-wave propagation   D. Skip propagation
The answer is D. Ionospheric propagation makes this possible. The sky wave that leaves the transmitter strikes the ionosphere and returns back to earth at a point, hundreds or even thousands of miles away. It is possible for the signal to be reflected from the earth and take a second hop, thereby increasing the range of communications.

Question 2C-2.1. What type of propagation involves radio signals that travel along the surface of the Earth? 11/12.
A. Sky-wave propagation   B. Knife-edge diffraction
C. E-layer propagation   D. Ground-wave propagation
The answer is D. The ground wave travels along the surface of the earth, gradually losing its strength until it is completely attenuated. See Figure 2C-1.1.

Question 2C-2.2. What is the meaning of the term ground wave propagation? 11/12.
A. Signals that travel along seismic fault lines
B. Signals that travel along the surface of the earth
C. Signals that are radiated from a ground-plane antenna
D. Signals that are radiated from a ground station to a satellite
The answer is B. See answer to question 2C-2.1.

Question 2C-2.3. Daytime communication on 3.725-MHz is probably via what kind of propagation when the stations are located a few miles apart but separated by a low hill blocking their line-of-sight path? 11/12.

A. Tropospheric ducting   B. Ground wave
C. Meteor scatter   D. Sporadic E
    The answer is B.

Question 2C-2.4. When compared to skip propagation, what is the usual effective range of ground wave propagation? 11/12.

A. Much smaller   B. Much greater
C. The same   D. Dependent on the weather
    The answer is A. Ground wave propagation is good up to 50 to 75 miles, whereas sky wave propagation is possible up to thousands of miles.

One (1) question must be from the following:

Question 2C-3.1. Why can a VHF or UHF radio signal that is transmitted toward a mountain often be received at some distant point in a different direction?

A. You can never tell what direction a radio wave is traveling in
B. These radio signals are easily reflected by objects in their path
C. These radio signals are easily bent by the ionosphere
D. These radio signals are sometimes scattered in the ectosphere
    The answer is B. Higher frequency signals tend to be reflected more than lower frequency signals. The term "VHF" indicates frequencies from 30 to 300 MHz. The term "UHF" indicates frequencies from 300 to 3000 MHz.

Question 2C-3.2. Why can the direction that a VHF or UHF radio signal is traveling be changed if there is a tall building in the way?

A. You can never tell what direction a radio wave is traveling in
B. These radio signals are easily reflected by objects in their path
C. These radio signals are easily bent by the ionosphere
D. These radio signals are sometimes scattered in the ectosphere
    The answer is B. See answer 2C-3.1.

Question 2C-4.1. What type of antenna polarization is normally used for communications on the 40-meter band?

A. Electrical polarization   B. Left-hand circular polarization
C. Horizontal polarization   D. Vertical polarization
    The answer is C. Most antennas in the 80 to 10 meter bands tend to have horizontal polarization. Some of the reasons for this are: (1) Man-made noise tends to be vertically polarized. A horizontal antenna will therefore pick up less of this noise than a vertical antenna. (2) Horizontal antennas, especially the wire type, are easier to construct. (3) Large antenna arrays are easier to construct in a horizontal system rather than a vertical system. (4) A horizontal antenna is

directional in a horizontal plane.

Question 2C-4.2. What type of antenna polarization is normally used for communications on the 80-meter band?
A. Right-hand circular polarization
B. Magnetic polarization
C. Horizontal polarization
D. Vertical polarization
　　The answer is C. See answer 2C-4.1.

Question 2C-4.3. What type of antenna polarization is normally used for communications on the 15-meter band?
A. Electrical polarization
B. Horizontal polarization
C. Right-hand circular polarization
D. Left-hand circular polarization
　　The answer is B. See answer 2C-4.1.

Question 2C-4.4. What type of antenna polarization is normally used for repeater communications on the 220-MHz band?
A. Vertical polarization
B. Horizontal polarization
C. Magnetic polarization
D. Left-hand circular polarization.
　　The answer is A. It is practical for repeaters to use vertical polarization since they communicate with mobile and portable stations that use vertical antennas. Vertical antennas produce vertical polarization.

Question 2C-4.5. What type of antenna polarization is normally used for repeater communications on the 1270-MHz band?
A. Enhanced polarization
B. Vertical polarization
C. Right-hand circular polarization
D. Left-hand circular polarization
　　The answer is B. See answer 2C-4.4.

Amateur Radio Practice

(4 questions)

One (1) question must be from the following:

Question 2D-1.1. How can an amateur station be protected against being operated by unauthorized persons? 13/14.
A. Install a carrier-operated relay in the main power line
B. Install a key-operated "ON/OFF" switch in the main power line
C. Post a "Danger -- High Voltage" sign in the station
D. Install ac line fuses in the main power line
   The answer is B. Some or all of the following measures should be taken to prevent the use of amateur station equipment by unauthorized personnel:
(1) Keep the amateur radio station room locked when the station is not in use.
(2) Lock the power switch when the radio station is not being used.
(3) Install a hidden power switch that must be turned on in order to operate the radio station.
(4) Post a sign in the station room, or on the door of the station room, indicating that only licensed authorized personnel may operate the radio station.
(5) Post the FCC licenses of those authorized to use the station in the station operating room.

Question 2D-2.1 Why should all antenna and rotor cables be grounded when an amateur station is not in use? 11/.
A. To lock the antenna system in one position
B. To avoid radio frequency interference
C. To save electricity
D. To protect the station and building from damage due to a nearby lightning strike
   The answer is D. By removing the antenna and rotor cables from the radio equipment and grounding them (the cables), the path of the lightning will be to **ground** and **not** to the station equipment.
   Before a storm, or when the station is not in use, the cables should be disconnected from the equipment, then removed from the home or radio shack, and connected to a good ground outside of the home. A good ground consists of one or more eight foot copper rods, buried in the earth near the base of the antenna tower.

Question 2D-2.2. How can an antenna system be protected from damage due to a nearby lightning strike? 11/.
A. Install a balun at the antenna feed point
B. Install an RF choke in the feed line
C. Ground all antennas when they are not in use

D. Install a line fuse in the antenna wire

The answer is C. The tower and the grounded parts of the antenna system should be permanently connected to a good ground at the base of the tower. During an electrical storm, or when the station is not in use, the antenna and rotor cables should be disconnected from the station equipment and connected to ground. See the answer to question 2D-2.1.

Question 2D-2.3. How can amateur station equipment be protected from damage due to lightning striking the electrical wiring in the building? 11/.
A. Use heavy insulation on the wiring
B. Keep the equipment on constantly
C. Disconnect the ground system
D. Disconnect all equipment after use, either by unplugging or by using a main disconnect switch

The answer is D. When the station is not in use, or during electrical storms, the AC power plugs of the equipment should be removed from their outlets. Also, the main power switch controlling the AC power to the station equipment should be shut down.

Question 2D-3.1. For proper protection from lightning strikes, what pieces of equipment should be grounded in an amateur station? 11/12.
A. The power supply primary
B. All station equipment
C. The feed line center conductors
D. The ac power mains

The answer is B. All metal cabinets housing the station equipment should be connected directly to a common point with large diameter, short length conductors. Another heavy conductor should connect this common point to the external ground described in answer 2D-2.1. The conductors that ground the equipment should be size #10 or larger.

Question 2D-3.2. What is a convenient indoor grounding point for an amateur station? 11/12.
A. A metallic cold water pipe   B. PVC plumbing
C. A window screen   D. A natural gas pipe

The answer is A. A metal water pipe can be used, provided it runs directly into the earth.

Question 2D-3.3. To protect against electrical shock hazards, to what should the chassis of each piece of equipment in an amateur station be connected? 11/12, App.6.
A. Insulated shock mounts   B. The antenna
C. A good ground connection   D. A circuit breaker

The answer is C. They should be connected to a common ground point with large diameter, short lengths of wire. The common ground point should be connected to a good ground, such as the one described in the answer to question 2D-2.1.

Question 2D-4.1. When climbing an antenna tower, what type of safety equipment should be worn? 11/11.
A. Grounding chain   B. A reflective vest
C. Long pants   D. A safety belt
    The answer is D.

Question 2D-4.2. For safety purposes, how high should all portions of a horizontal wire antenna be located?
A. High enough so that a person cannot touch them from the ground
B. Higher than chest level
C. Above knee level
D. Above electrical lines
    The answer is A. The horizontal antenna wire should be placed as high as is practical with proper physical supports. Do not use thin tree branches for support, or else, the antenna will be down with the first snow or ice storm. It is important to avoid placing the antenna under electric wires.
    The antenna should be high enough so that a person on the ground cannot touch it. Touching an antenna during high power transmissions can result in electric shock or RF burns.

Question 2D-4.3. While assisting another person working on an antenna tower what type of safety equipment should a person on the ground wear?
A. A reflective vest   B. A safety belt
C. A grounding chain   D. A hard hat
    The answer is D. He should wear a helmet (hard hat) to prevent head injuries in the event that any tools or parts are accidentally dropped.

One (1) question must be from the following:

Question 2D-5.1. What is a likely indication that radio frequency interference to a receiver is caused by front-end overload?
A. A low pass filter at the transmitter reduces interference sharply
B. The interference is independent of frequency
C. A high pass filter at the receiver reduces interference little or not at all
D. Grounding the receiver makes the problem worse
    The answer is B. The interference covers a wide range of frequencies. In the case of TV overload, it causes image distortion or destruction to most or all channels. See the answer to question 2D-5.4.

Question 2D-5.2. What is likely the problem when radio frequency interference occurs to a receiver regardless of frequency, while an amateur station is transmitting?
A. Inadequate transmitter harmonic suppression
B. Receiver VR tube discharge

C. Receiver overload
D. Incorrect antenna length
   The answer is C. See answers to questions 2D-5.1 and 2D-5.4.

Question 2D-5.3. What type of filter should be installed on a TV receiver tuner as the first step in preventing overload from an amateur station transmission? 9/10.
A. Low pass   B. High pass   C. Band pass   D. Notch
   The answer is B. The first step would be to install a high-pass filter at the TV receiver's input. The high-pass filter will pass the high frequency TV signals, but will block the low frequency transmitter signal. It should be installed as close to the TV tuner as possible.

Question 2D-5.4. What is meant by receiver overload? 9/10.
A. Interference caused by transmitter harmonics
B. Interference caused by overcrowded band conditions
C. Interference caused by strong signals from a nearby transmitter
D. Interference caused by turning the receiver volume too high
   The answer is C. Receiver overload refers to a strong fundamental signal from a transmitter that gets into a receiver's front end. Because of its strength, it swamps the receiver's input circuits, which lack the selectivity to suppress the strong signal. The strong interfering signal will generally cause interference over a broad frequency range.

Question 2D-6.1. What is meant by harmonic radiation? 9/10.
A. Transmission of signals at whole number multiples of the funda-
   mental (desired) frequency
B. Transmission of signals that include a superimposed 60-Hz hum
C. Transmission of signals caused by sympathetic vibrations from a
   nearby transmitter
D. Transmission of signals to produce a stimulated emission in the air
   to enhance skip propagation
   The answer is A. A harmonic is a whole numbered multiple of a fundamental frequency. For instance, if the fundamental frequency is 4.0 MHz, its second harmonic is 8.0 MHz, its third harmonic is 12.0 MHz, etc. A transmitter, producing a fundamental frequency signal, will also produce harmonics which will be radiated along with the fundamental. The harmonics will cause interference to other services unless steps are taken to prevent their radiation.

Question 2D-6.2. Why is harmonic radiation by an amateur station undesirable? 9/10.
A. It will cause interference to other stations and may result in
   out-of-band signal radiation
B. It uses large amounts of electric power
C. It will cause sympathetic vibrations in nearby transmitters
D. It will produce stimulated emission in the air above the transmit-

ter, thus causing aurora

The answer is A. Harmonic radiation is undesirable because it may cause interference to other services. For instance, if an amateur is operating at 29 MHz, the second harmonic is 58 MHz. If a 58 MHz harmonic is radiated from the transmitter, it would interfere with TV channel 2 which occupies the band of frequencies from 54 to 60 MHz.

Question 2D-6.3. What type of interference may radiate from a multi-band antenna connected to an improperly tuned transmitter?
A. Harmonic radiation   B. Auroral Distortion
C. Parasitic excitation   D. Intermodulation

The answer is A. It will radiate harmonic interference. This is because a multi-band antenna is designed to operate at several frequencies that are harmonically related. An antenna tuner and/or a low pass filter should be used when using a multi-band antenna. The antenna tuner has tuned circuits that select the desired signal and reject other signals. The low pass filter described in the answer to question 2D-6.6, will prevent harmonics of the desired signal from getting out.

Question 2D-6.4. What is the purpose of shielding in a transmitter? 9/10.
A. It gives the low pass filter structural stability
B. It enhances the microphonic tendencies of radiotelephone transmitters
C. It prevents unwanted RF radiation
D. It helps maintain a sufficiently high operating temperature in circuit components

The answer is C. Shielding the transmitter prevents the emission of any RF signals, spurious, or otherwise, from anywhere except the transmission line that feeds the antenna. If we then place a low pass filter between the transmitter and the transmission line, we will complete the suppression of unwanted radiation from the transmitter.

The various cabinets and housings that shield the transmitter and other station accessories, should be connected to a common ground point and then to a good ground to be effective.

Question 2D-6.5. What is the likely problem when interference is observed on only one or two channels of a TV receiver while an amateur station is transmitting? 9/10.
A. Excessive low-pass filtering   B. Sporadic E de-ionization
C. Receiver front-end overload   D. Harmonic radiation

The answer is D. The chances are that the interference is caused by harmonic radiation from the transmitter. The most important step in eliminating this type of interference is to install a low pass filter at the transmitter. We should also shield the transmitter and ground all cabinets and accessory housings.

Question 2D-6.6. What type of filter should be installed on an

amateur transmitter as the first step in reducing harmonic radiation? 9/10.

A. Key click filter   B. Low pass filter
C. High pass filter   D. CW filter

The answer is B. A low pass filter should be installed at the transmitter. It will reduce the transmission of harmonics. A low pass filter is a device made of coils and capacitors. It passes signals whose frequency is below a certain point, called the "cut-off" frequency. It prevents the passage of signals above the cut-off frequency. These signals, above the cut-off frequency, are the undesired harmonics.

One (1) question must be from the following:

Question 2D-7.1. Why should the impedance of a transmitter final-amplifier circuit match the impedance of the antenna or feedline? 11/12.
A. To prevent sympathetic vibrations in nearby radio equipment
B. To obtain maximum power transfer to the antenna
C. To help maintain a sufficiently high operating temperature in circuit components
D. To create a maximum number of standing waves on the feed line

The answer is B. The impedance of the transmitter final amplifier circuit must match (be equal to) the impedance of the antenna or feedline in order to transfer maximum RF power from the transmitter to the antenna or feedline. Matching impedances also cuts down on the radiation of spurious and harmonic signals. If a feedline (transmission line) is used, it is most important for the transmission line impedance to be equal to the impedance of the antenna at its feedpoint.

Question 2D-7.2. What is the term for the measurement of the impedance match between a transmitter final-amplifier circuit and the antenna or feedline? 11/12.
A. Voltage flyback ratio   B. Impedance sine ratio
C. Standing wave ratio   D. Current over-feed ratio

The answer is C. The **Standing Wave Ratio (SWR)** is the term that describes the measurement of the impedance match between the transmitter and the antenna or feedline. The formula for Standing Wave Ratio is:

$$SWR = \frac{Z_L}{Z_{TL}}$$

where: Z_L is the impedance of the load and Z_{TL} is the impedance of the transmission line

An SWR of 1 to 1 indicates that the impedances are matched and maximum RF is being transferred between the transmitter and the antenna or feedline. If the impedances are not matched, some of the RF power will be reflected back to the transmitter from the antenna. This will result in an SWR that is greater than 1 to 1, such as 2 to 1 or 3 to 1, etc.

Question 2D-7.3. What accessory is used to measure RF power being reflected back down the feedline from the antenna to the transmitter? 11/12.
A. An SWR meter   B. RF tuner
C. S-meter   D. Field strength meter
   The answer is A. An SWR bridge (called a Reflectometer) reads forward and reflected voltages, and can display the value of the SWR on the face of the meter.

Question 2D-7.4. What accessory is often used to measure voltage standing wave ratio? 11/12.
A. Ohmmeter   B. Ammeter   C. SWR bridge   D. Current bridge
   The answer is C. Voltage standing wave ratio (VSWR) and standing wave ratio (SWR) are terms that are used interchangeably and mean the same thing. An SWR bridge or Reflectometer is used to measure VSWR. See answers to questions 2D-7.2 and 2D-7.3.

Question 2D-7.5. Where should a standing wave ratio bridge be connected to indicate the impedance match of a transmitter and an antenna?
A. Between the antenna and matchbox
B. Between the key and transmitter
C. Between the mike and transmitter
D. Between the transmitter and matchbox
   The answer is D. It should be inserted in the transmission line that connects the transmitter to the antenna. However, since we are most interested in the maximum transfer of RF power directly into the antenna, we should place the SWR bridge right at the point where the transmission line feeds the antenna. This, of course, may not be practical, so we use the next best procedure which is to place it between the transmitter and the transmission line.

Question 2D-7.6. Coaxial feedlines should be operated with what kind of standing wave ratio?
A. As high as possible
B. As low as possible
C. Standing wave ratio is not important
D. Standing wave ratios cannot be measured in a coaxial cable
   The answer is B. Most coaxial cables have an impedance of from 50 to 75 ohms. This matches the center impedance of a half-wave dipole and will produce a low SWR. Also, the outer shield of the coaxial cable prevents interaction between the cable and nearby metal objects, which, in turn, keeps the SWR down.

Question 2D-7.7. If the standing wave ratio bridge reading is higher at 3700-kHz than at 3750-kHz, what does this indicate about the antenna?

A. Too long for optimal operation at 3700 kHz
B. Broadbanded
C. Good only for 37-meter operation
D. Too short for optimal operation at 3700 kHz
The answer is D. This indicates that the length of the antenna is such that its resonant frequency is closer to 3,750 kHz than it is to 3,700 kHz. The SWR is lowest when the antenna is resonant to the frequency being fed to it. Since 3,750 kHz has a lower wavelength than 3,700 kHz, the antenna length, which varies directly with wavelength, would be too short at 3,700 kHz.

Question 2D-7.8. If the standing wave ratio bridge reading is lower at 3700-kHz than at 3750-kHz, what does this indicate about the antenna?
A. Too long for optimal operation at 3750 kHz
B. Broadbanded
C. Good only for 37-meter operation
D. Too short for optimal operation at 3750 kHz
The answer is A. This indicates that the antenna's resonant frequency is closer to 3,700 kHz than it is to 3,750 kHz. Since 3,700 kHz has a longer wavelength than 3,750 kHz, the antenna length is too long for 3,750 kHz. See answer 2D-7.7.

Question 2D-8.1. What kind of standing wave ratio bridge reading may indicate poor electrical contact between parts of an antenna system?
A. An erratic reading   B. An unusually low reading
C. No reading at all   D. A negative reading
The answer is A. A high SWR reading will indicate poor electrical contact between parts of an antenna system. If an antenna system has been operating properly with a low SWR reading for a period of time and suddenly develops a high SWR reading, one should suspect corrosion at the contacts, or poor electrical contact at other points in the antenna system. If the electrical contact is intermittent, the SWR reading will be erratic.

Question 2D-8.2. High standing wave ratio bridge readings measured from a half-wave dipole antenna being fed by coaxial cable can be lowered by doing what to the antenna?
A. Change the electrical length of the antenna
B. Reduce the diameter of the antenna's radiating element
C. Connect a short jumper wire across the antenna's center insulator
D. Use a feed line having less loss per foot
The answer is A. Assuming that it has been determined that the antenna itself is at fault, we can reduce the SWR reading by changing the length of the antenna. We can first reduce the length by a few inches and measure the SWR reading. If it is lower, we continue to reduce the length until we have minimum reading. If the first reduction in length gave us a higher reading, it indicated that the antenna was too short. We therefore increase the length of the

antenna, a few inches at a time, until the SWR reading is minimum.

One (1) question must be from the following:

Question 2D-9.1. What precautions should you take when working with a 1270-MHz waveguide?
A. Make sure that the RF leakage filters are installed at both ends of the waveguide
B. Never look into the open end of a waveguide when RF is applied
C. Minimize the standing-wave ratio before you test the waveguide
D. Never have both ends of the waveguide open at once when RF is applied

The answer is B. A waveguide is a transmission line that conducts high frequency energy from one point to another, in much the same manner as a transmission line. A waveguide consists of a rectangular or circular conducting tube through which the energy is transmitted in the form of electromagnetic waves. See Figure 2D-9.1. These waves are both electric and magnetic, and they are propagated through the waveguide tubes by reflections against the inner walls. Skin effect prevents the electromagnetic energy from escaping through the metal walls.

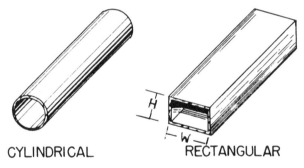

CYLINDRICAL   RECTANGULAR

Fig. 2D-9.1. Basic forms of waveguides.

The primary reason for using waveguides at high frequencies is that they have low losses. At high frequencies, coaxial cable exhibits very high dielectric losses. Waveguides use air as the dielectric, making the dielectric losses almost negligible.

The reason that you should not look into the open end of a waveguide, when RF is applied to it, is that the RF energy can be harmful to your eyes. See answer 2D-9.2.

Question 2D-9.2. What precautions should you take when you mount a VHF or UHF antenna in a permanent location?
A. Make sure that no one can be near the antenna when you are transmitting
B. Make sure that the RF shield screens are in place
C. Make sure that the antenna is near the ground to maximize directional effect

D. Make sure you connect an RF leakage filter at the antenna feed point

The answer is A. When RF energy strikes the human body, it causes internal molecular agitation, which in turn, causes heat. The heat can damage the body tissues. The amount of possible damage depends upon the frequency of the RF, the amount of RF power, the type of RF signal, the cumulative amount of exposure and the part of the body that is involved.

As the frequency increases, the RF damage to the body increases. Below 10 MHz, there is minimal effect. At 27 MHz, there is a warming effect on the bones. Continuous exposure to high power VHF (30 to 300 MHz) and UHF (300 to 3000 MHz) radiation, can cause considerable damage to body tissues.

The higher the RF power, the greater are the biological effects. In addition to other reasons, this is one of the most important reasons to use low power. An amateur operator should always use the least amount of power that insures reliable communications. Low power and a good antenna give much better results than a poor antenna with high power.

Fortunately, most amateurs use low power and produce intermittent radiation. More than half of the operating time is spent looking and listening, rather than transmitting. While transmitting, the signal is not always maximum. In CW, there is no RF output a good deal of the time. In single sideband, the peak RF output is only reached a fraction of the time. However, in the case of FM and RTTY, the signal is at peak practically all the time. This fact should be taken into consideration when planning the construction of the station and antenna system.

Different parts of the body react differently to RF. The eyes, testes and ovaries are most susceptible to injury from RF radiation.

The above paragraphs should not cause concern or alarm to prospective amateurs. If common sense is used and safety precautions are observed, an amateur can enjoy a safe hobby for a long, long time. The following safety measures should be taken to insure minimum exposure to RF:

1. Locate the antenna as high as practical and as far away from peoples' living quarters as possible. A directional antenna or dish type antenna should not be pointed in the direction of people.

2. Use good quality coaxial cable. The transmission line should not radiate RF. The antenna should do all the radiating.

3. Use the lowest amount of power to maintain reliable communications.

4. All equipment should be encased in properly shielded, grounded metal cabinets.

5. A good external ground system should be constructed near the antenna system. The grounded cabinets should be connected to this ground.

6. Make certain that the antenna impedance, the transmission line impedance, and the transmitter output impedance are all matched and

that the SWR is low.

7. Do not work on the antenna or inside of the transmitter unless you are 100% certain that no one can turn the power on.

8. When working with a hand-held transmitter, keep the antenna away from your head, especially away from your eyes.

9. Do not look into a "live" waveguide. The VHF or UHF energy can be quite harmful to your eyes.

10. If possible, use double shielding on transmission lines and connectors. We want to minimize RF leakage.

11. We must again stress that there should be no cause for alarm or concern if common sense is used with regard to safety measures. Ham Radio is a safe hobby!

Question 2D-9.3. What precautions should you take before removing the shielding on a VHF or UHF power amplifier?

A. Make sure all RF screens are in place at the antenna
B. Make sure the feed line is properly grounded
C. Make sure the amplifier cannot be accidentally energized
D. Make sure that the RF leakage filters are connected
    The answer is C. See answer 2D-9.2.

Question 2D-9.4. Why should you use only good-quality, well-constructed coaxial cable and connectors for a VHF or UHF antenna system?

A. To minimize RF leakage
B. To reduce parasitic oscillations
C. To maximize the directional characteristics of your antenna
D. To maximize the standing-wave ratio of the antenna system
    The answer is A. See answer 2D-9.2.

Question 2D-9.5. Why should you be careful to position the antenna of your 220-MHz hand-held transceiver away from your head when you are transmitting?

A. To take advantage of the directional effect
B. To minimize RF exposure
C. To use your body to reflect the signal, improving the directional characteristics of the antenna
D. To minimize static discharges
    The answer is B. The eyes are especially sensitive to RF radiation. See answer 2D-9.2.

Question 2D-9.6. How can you minimize RF exposure when you are operating your 220-MHz hand-held transceiver?

A. Position the antenna near the ground
B. Use a shielded RF screen around your antenna
C. Use a special short "stubby duck" antenna
D. Position the antenna away from your head
    The answer is D. See answers 2D-9.5 and 2D-9.2.

Question 2D-9.7. Why should you be careful to position the antenna of your 1270-MHz hand-held transceiver away from your head when you are transmitting?
A. To take advantage of the directional effect
B. To use your body to reflect the signal, improving the directional characteristics of the antenna
C. To minimize static discharges
D. To minimize RF exposure
    The answer is D. See answers 2D-9.5 and 2D-9.2.

Question 2D-9.8. How can you minimize RF exposure when you are operating your 1270-MHz hand-held transceiver?
A. Position the antenna near the ground
B. Use a shielded RF screen around your antenna
C. Use a special short "stubby duck" antenna
D. Position the antenna away from your head
    The answer is D. See answers 2D-9.5 and 2D-9.2.

Question 2D-9.9. How can you minimize RF leakage from your VHF or UHF antenna system?
A. Use open-wire line for antenna feed line
B. Use only good-quality, well-constructed coaxial cable and connectors
C. Use specially-shielded AC line cords with all your equipment
D. Use an RF leakage filter on the antenna feed line
    The answer is B. The outer conductors of the cable and connectors act as shields and can minimize RF leakage. See answer 2D-9.2.

Question 2D-9.10. Why should you make sure your VHF or UHF amplifier cannot be energized before you open the amplifier enclosure?
A. To minimize static discharge when you open the enclosure
B. To minimize RF exposure and prevent electric shock
C. To minimize the effects of hand capacitance
D. To prevent exposure to Cerenkov radiation from the amplifier
    The answer is B. See answer 2D-9.2.

Question 2D-9.11. Why should you never look into a VHF or UHF waveguide when RF is applied?
A. Because the fluorescent coating inside the waveguide gets very bright
B. Because exposure to VHF or UHF RF energy can be harmful to your eyes
C. Because the waveguide might not be properly grounded
D. Because the Cerenkov Effect may scatter RF energy
    The answer is B. The eyes are especially sensitive to RF radiation. See answer 2D-9.2.

Question 2D-9.12. Why should you be sure that your transmitter

cannot be energized before you work on your VHF or UHF antennas?
A. Because operating the transmitter when the antennas are disconnected might harm the transmitter
B. Because exposure to VHF or UHF RF energy can be harmful
C. Because if the transmitter is operated while you are touching the antenna, the radiated energy might be out of an amateur band
D. Because accidental operation might blow a fuse

The answer is B. Touching a live VHF or UHF antenna may cause RF burns. See answer 2D-9.2.

SUBELEMENT 2E
Electrical Principles
(4 questions)

One (1) question must be from the following:

**Question 2E-1.1. Electrons will flow in a copper wire when its two
ends are connected to the poles of what kind of source? 1/1**
A. Electromotive or voltage   B. Donor
C. Reactive   D. Resistive
    The answer is A. Electrons will flow in the wire when its two
ends are connected to a voltage source. Voltage is the electrical
pressure that forces electrons to flow through a wire or circuit.
Another term for voltage is **ELECTROMOTIVE FORCE.**
    A typical example of a voltage source is the common dry cell that
is used in a flashlight or calculator.

**Question 2E-1.2. The pressure in a water pipe is comparable to what
force in an electrical circuit? 1/1.**
A. Current   B. Resistive
C. Gravitational   D. Voltage
    The answer is D. The pressure in a water system pushes water in
a manner similar to the voltage "pushing" electrons.

Question 2E-1.3. What are the two polarities of a voltage? 1/1.
A. Right-hand and left-hand
B. Forward and reverse
C. Positive and negative
D. Clockwise and counter clockwise
    The answer is C. A voltage has a positive (+) and a negative (-)
polarity. A battery is a source of voltage. Figure 2E-2.1 illustrates
a battery hooked up to a bulb. Note the positive and negative
terminals of the battery. The electrons leave the battery at its
negative terminal and return at its positive terminal.

Fig. 2E-2.1. A battery hooked up to a bulb.

FCC omitted question 2E-2.1 when they made up the question pool.

Question 2E-2.2. What type of current changes direction over and over again in a cyclical manner? 3/3.
A. Direct current   B. Alternating current
C. Negative current   D. Positive current
  The answer is B. Alternating current changes its direction of flow many, many times in one second. First it flows in one direction, then it reverses itself and flows in the other direction. The current in our homes is alternating current. It reverses itself 60 times per second.

Question 2E-2.3. What is a type of electrical current called that does not periodically reverse direction? 1/1.
A. Alternating current   B. Periodic current
C. Direct current   D. Positive current
  The answer is C. Direct current flows in one direction only. A battery is a source of direct current. Figure 2E-2.3 shows the direction of current flow in a simple DC circuit.

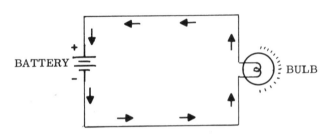

Fig. 2E-2.3 A source of direct current.

Question 2E-3.1. List at least four good electrical insulating materials. 1/1.
A. Glass, air, plastic, porcelain
B. Glass, wood, copper, porcelain
C. Paper, glass, air, aluminum
D. Plastic, rubber, wood, carbon
  The answer is A. Wood, silk, glass and bakelite are other good electrical insulators. Insulators will not permit current to flow through them.

Question 2E-3.2. List at least three good electrical conductors.
A. Copper, gold, mica   B. Gold, silver, wood
C. Gold, silver, aluminum   D. Copper, aluminum, paper
  The answer is C. Copper is also an excellent conductor. Conductors also allow current to flow through them easily.

Question 2E-3.3. What is the term for the lowest voltage that will cause a current in an insulator?
A. Avalanche voltage   B. Plate voltage
C. Breakdown voltage   D. Zener voltage

The answer is C. If we place an insulator across a voltage source, there will normally be NO current flow. However, if we continue to increase the voltage to a large value, a point is reached when the insulation properties of the insulator break down and current flows. The voltage at this point is called the "breakdown voltage". When the current flows, the insulator becomes a conductor.

The breakdown voltage varies with different insulators.

One (1) question must be from the following:

Question 2E-4.1. What is the term for a failure in an electrical circuit that causes excessively high current? 1/1.
A. Open circuit   B. Dead circuit
C. Closed circuit   D. Short circuit

The answer is D. A short circuit presents very little resistance to a voltage source, and causes a heavy, damaging current to flow. Figure 2E-4.1 illustrates a short circuit. Note that the current flows through the short and not through the bulb.

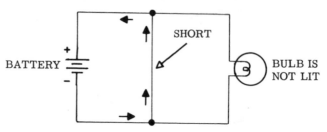

Figure. 2E-4.1. A short circuit.

Question 2E-4.2. What is the term for an electrical circuit in which there can be no current flow? 1/1.
A. A closed circuit   B. A short circuit
C. An open circuit   D. A hyper circuit

The answer is C. An open circuit indicates a break or opening in the circuit. Current cannot flow unless the circuit is complete. Figure 2E-4.2 illustrates an open circuit.

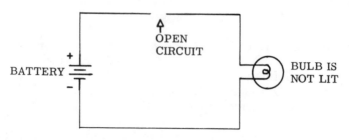

Fig. 2E-4.2. An open circuit.

Question 2E-5.1. What is consumed when a voltage is applied to a

circuit causing an electrical current to flow? 16/19.
A. Energy   B. Volts   C. Amps   D. Electrons
   The answer is A. When an electrical current flows through a circuit, electrical energy is transformed into another form of energy. If the circuit is resistive, the electrical energy is transformed into heat energy. The rate at which the energy is transformed is called POWER.

Question 2E-6.1. What is the approximate length, in meters, of a radio wave having a frequency of 3.725-MHz? 11/12.
A. 160 meters   B. 80 meters   C. 40 meters   D. 30 meters
   The answer is B. We solve this by using the following formula for wavelength:

$$\text{Wavelength, in meters} = \frac{300}{\text{frequency in MHz}} =$$

$$\frac{300}{3.725} = 80.5 \text{ meters}$$

   Simply stated, the wavelength in meters is equal to 300 divided by the frequency in MegaHertz.

Question 2E-6.2. What is the relationship between frequency and wavelength? 11/12.
A. As frequency increases, wavelength decreases
B. As frequency increases, wavelength increases
C. Frequency and wavelength are not related
D. As frequency decreases, wavelength decreases
   The answer is A. Two of the factors used to describe an alternating current signal are its frequency and its wavelength. They are inversely related. As one goes up, the other goes down. A high frequency has a small wavelength; a low frequency has a large wavelength. If we divide 300 by the frequency in MegaHertz (MHz), we get the wavelength in meters. On the other hand, if we divide 300 by the wavelength in meters, we get the frequency in MHz. See answer to Question 2E-6.1. If we are dealing with frequency in kiloHertz (kHz), we use 300,000 instead of 300.

Question 2E-6.3. What is the approximate length, in meters, of a radio wave having a frequency of 21.120-MHz? 11/12.
A. 80 meters   B. 40 meters   C. 15 meters   D. 10 meters
   The answer is C. We find this by using the formula in answer 2E-6.1.

$$\frac{300}{21.120 \text{ MHz}} = 14.2 \text{ meters}$$

One (1) question must be from the following:

Question 2E-7.1. What is the difference between radio frequencies and audio frequencies? 3/3.

A. Audio frequencies are those below 20,000 Hz and radio frequencies are those above 20,000 Hz
B. Audio frequencies are those below 50,000 Hz and radio frequencies are those above 50,000 Hz
C. Audio frequencies are those below 10,000 Hz and radio frequencies are those above 10,000 Hz
D. Audio frequencies are those above 20,000 Hz and radio frequencies are those below 20,000 Hz

The answer A. Radio frequencies are higher than audio frequencies. Audio frequencies are generally between 16 Hertz (Hz) and 16,000 Hz. Radio frequencies are above 20,000 Hz.

Question 2E-7.2. What type of frequency is 3,500,000 Hertz?   3/3.
A. An audio frequency   B. A microwave frequency
C. An intermediate frequency   D. A radio frequency
The answer is D. See answer 2E-7.1.

Question 2E-7.3. Radio frequencies are those above what frequency? 3/3.
A. 20 Hz   B. 2000 Hz   C. 20,000 Hz   D. 2,000,000 Hz
The answer is C. See answer 2E-7.1.

Question 2E-8.1. What type of frequency is 350 Hertz? 3/3.
A. An audio frequency   B. A microwave frequency
C. An intermediate frequency   D. A radio frequency
The answer is A. See answer 2E-7.1.

Question 2E-8.2. Audio frequencies are those below what frequency? 3/3.
A. 10 Hz   B. 20 Hz   C. 10,000 Hz   D. 20,000 Hz
The answer is D. See answer 2E-7.1.

Question 2E-8.3. What type of frequency is 3,500 Hertz? 3/3.
A. Audio frequency   B. Radio frequency
C. Hyper-frequency   D. Super-high frequency
The answer is A. See answer 2E-7.1.

Question 2E-9.1. What is the unit of electromotive force? 1/1.
A. Ampere   B. Volt   C. Ohm   D. Watt
The answer is B. A kilovolt is one thousand volts. A millivolt is one thousandth of a volt. A microvolt is one millionth of a volt.

Question 2E-10.1. What is the unit of electrical current? 1/1.
A. Volt   B. Watt   C. Ampere   D. Ohm
The answer is C. A milliampere is one thousandth of an ampere. A microampere is one millionth of an ampere.

Question 2E-11.1. What is the unit of electrical power? 1/1.
A. Ohm   B. Watt   C. Volt   D. Ampere

The answer is B. A kilowatt is one thousand watts. A milliwatt is one thousandth of a watt.

One (1) question must be from the following:

Question 2E-12.1. What is a Hertz? 3/3.
A. A unit of measure of current
B. A unit of measure of capacitance
C. A unit of measure of frequency
D. A unit of measure of power
    The answer is C. A kiloHertz (kHz) is one thousand Hertz. A megaHertz (MHz) is one million Hertz.

Question 2E-12.2. What is another popular term for Hertz? 3/3.
A. Cycles per second   B. Frequency per wavelength
C. Wavelength per cycle   D. Meters per frequency
    The answer is A. The original term for Hertz was "cycles per second", or simply, "cycles".

Question 2E-13.1. A frequency of 40,000 Hertz is equal to how many kiloHertz? 3/3.
A. 40   B. 4   C. 400   D. 0.04
    The answer is A. Hertz is changed to kiloHertz by dividing by 1,000, or by simply moving the decimal point three places to the left.

$$\frac{40,000 \text{ Hertz}}{1,000} = 40.0 \text{ kiloHertz}$$

Question 2E-13.2. A current of 20 millionths of an ampere is equal to how many microamperes? 1/1.
A. 0.2   B. 2   C. 20   D. 200
    The answer is C. One microampere IS a millionth of an ampere (See answer 2E-10.1). Therefore, 20 millionths of an ampere are equal to 20 microamperes.

Question 2E-13.3. A current of 2000 milliamperes is equivalent to how many amperes? 1/1.
A. 0.002 A   B. 0.2 A   C. 2 A   D. 2000 A
    The answer is C. An ampere is equal to 1,000 milliamperes. We change milliamperes to amperes by dividing the milliamperes by 1,000, or simply by moving the decimal point three places to the left.

$$\frac{2,000 \text{ milliamperes}}{1,000} = 2.0 \text{ amperes}$$

Question 2E-13.4. What do the prefixes mega- and centi- mean? 1,3/1,12.
A. 1,000,000 and 0.01   B. 0.001 and 0.01
C. 1,000,000 and 100   D. 0.001 and 100
    The answer is A. "Meg" or "mega" means 1,000,000 times. One MegaHertz is equal to 1,000,000 Hertz. One megohm is equal to

1,000,000 ohms.

"Centi" means one hundredth. One centimeter is one hundredth of a meter.

Question 2E-13.5. What do the prefixes <u>micro-</u> and <u>pico-</u> mean? 1,3/1,3.

A. 1,000,000 and 1,000   B. 1,000,000 and 1,000,000,000
C. 0.000,001 and 0.001   D. 0.000,001 and 0.000,000,000,001

The answer is D. "Micro" means one millionth. A microvolt is one millionth of a volt.

"Pico" means a millionth of a millionth or a trillionth. One picofarad (pf) is equal to one trillionth of a farad (f).

$$1 \text{ pf} = \frac{1}{1,000,000,000,000} \text{ f}$$

Question 2E-13.6. Your receiver dial is calibrated in megahertz and shows a signal at 1200 MHz. At what frequency would a dial calibrated in gigahertz show the signal?

A. 1.2 GHz   B. 12 GHz   C. 120 GHz   D. 1200 GHz

The answer is A. A gigahertz (GHz) is equal to 1 billion Hz or 1000 MHz. Therefore, in order to change megahertz to gigahertz, we divide by 1000. 1200 MHz ÷ 1000 = 1.2 GHz.

Question 2E-13.7. Your receiver dial is calibrated in gigahertz and shows a signal at 1.27 GHz. At what frequency would a dial calibrated in megahertz show the signal?

A. 1.27 MHz   B. 12.7 MHz   C. 127 MHz   D. 1270 MHz

The answer is D. In order to change GHz to MHz, we multiply by 1000. Therefore, 1.27 x 1,000 = 1270 MHz.

Question 2E-13.8. Your receiver dial is calibrated in megahertz and shows a signal at 223.9 MHz. At what frequency would a dial calibrated in kilohertz show the signal?

A. 0.223 kHz   B. 2239 kHz   C. 22,390 kHz   D. 223,900 kHz

The answer is D. In order to change MHz to kHz, we multiply by 1000. Therefore, 223.9 x 1000 = 223,900 kHz.

SUBELEMENT 2F
Circuit Components
(2 questions)

One (1) question must be from the following:

Question 2F-1.1. What is the general relationship between the thickness of a quartz crystal and its fundamental operating frequency? /9.
A. The thickness of a crystal does not affect operating frequency
B. Thinner crystals oscillate at lower frequencies
C. Thinner crystals oscillate at higher frequencies
D. Thicker crystals oscillate at higher frequencies
 The answer is C. The thickness and fundamental frequency vary inversely with each other. The thinner the crystal, the higher is its fundamental operating frequency; the thicker the crystal, the lower is its fundamental frequency.

Question 2F-1.2. What is the schematic symbol for a quartz crystal? 8/9.

The answer is A.

Question 2F-1.3. What chief advantage does a crystal controlled transmitter have over one controlled by a variable frequency oscillator? 8/9.
A. The crystal-controlled transmitter will not produce key clicks
B. The crystal-controlled transmitter has better frequency stability
C. The crystal-controlled transmitter does not need to be tuned
D. The crystal-controlled transmitter can operate at higher power output
 The answer is B. The frequency stability of a transmitter is its ability to maintain a constant frequency, regardless of changes in voltage, temperature or other factors.

Question 2F-2.1. What two internal components of a D'Arsonval meter interact to cause the indicating needle to move when current flows through the meter? 2/2.
A. A diode and a capacitor
B. A transformer and a resistor
C. A coil of wire and a permanent magnet
D. A dipole and a balun
 The answer is C. A D'Arsonval meter is the basic meter movement used in analog type meters. (An analog meter has a scale with a

moving pointer, as opposed to a digital meter which has a read-out). Figure 2F-2.1 illustrates a D'Arsonval meter movement.

The magnetic field of the movable coil (that carries the current being measured), interacts with the magnetic field of the permanent magnet, causing the movable coil, with its attached indicating needle, to move.

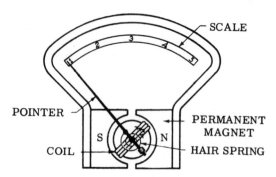

Fig. 2F-2.1 The D'Arsonval Meter Movement.

Question 2F-2.2. What does a voltmeter measure? 1/2.
A. Resistance   B. Current   C. Power   D. Voltage
The answer is D.

One (1) question must be from the following:

Question 2F-3.1. Draw the schematic diagram of a triode vacuum tube and label the elements. 6/6.
See Figure 2F-3.1.

Question 2F-3.2. Draw the schematic symbol for a tetrode vacuum tube and label the elements. 6/6.
See Figure 2F-3.2.

Question 2F-3.3. Draw the schematic symbol for a pentode vacuum tube and label the elements. 6/6.
See Figure 2F-3.3.

Question 2F-4.1. What device should be included in electronic equipment to protect it from damage resulting from a short circuit? 1/1.
A. Fuse   B. Tube   C. Transformer D. Filter
The answer is A. A fuse is placed in series with the circuit. When an excessive amount of current flows through the circuit, it also flows through the fuse. The fuse element heats up and burns out. This causes the circuit to be open. The current flow in the entire circuit drops to zero and the electronic equipment is protected.

Question 2F-4.2. What happens to a fuse when an excessive amount of

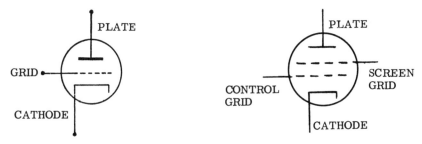

Fig. 2F-3.1. Schematic symbol   Fig. 2F-3.2. Schematic symbol of
of a triode vacuum tube.   a tetrode vacuum tube.

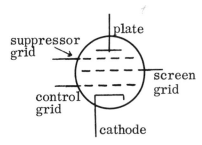

Fig. 2F-3.3. Schematic symbol of a pentode
vacuum tube.

current flows through it? 1/1.
A. The fuse explodes, the circuit is destroyed, the current increases
B. The fuse glows red or orange, the circuit shorts, the current increases
C. The fuse melts, the circuit shorts, the current increases
D. The metal conductor inside the fuse melts, the circuit opens, the current stops
The answer is D. See answer 2F-4.1.

SUBELEMENT 2G
Practical Circuits
(2 questions)

One (1) question must be from the following:

Question 2G–1.1. Draw a block diagram representing the stages in a simple crystal-controlled telegraphy transmitter. 8/9, 10.
See Figure 2G–1.1. The crystal-controlled oscillator generates a single fixed frequency. If we wish to change the frequency, we must change the crystal to one of a different frequency. The RF output of the oscillator is then fed to an RF amplifier, which amplifies the RF output of the oscillator. The RF amplifier may also serve as a frequency multiplier. The frequency multiplier multiplies the frequency of the signal up to the desired frequency. The power amplifier takes the signal from the RF amplifier and amplifies it in terms of power. The signal is then fed to the antenna where it is radiated. The RF amplifier also acts as a buffer between the power amplifier and the oscillator. It prevents changes in the power amplifier from affecting the oscillator. This improves the stability of the transmitter.

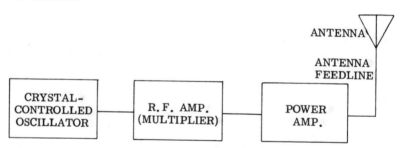

Fig. 2G–1.1. A crystal-controlled transmitter.

A transmitter need not have three stages. The oscillator alone can act as a transmitter. However, the more stages the transmitter has, the better it is.

Question 2G–1.2. What type of transmitter does this block diagram (Figure 2G–1.2) represent?

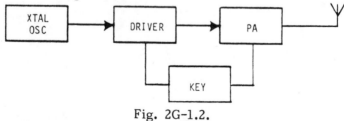

Fig. 2G–1.2.

A. A simple crystal-controlled receiver
B. A simple crystal-controlled transmitter
C. A simple sideband transmitter
D. A VFO controlled transmitter

The answer is B. The diagram is similar to the diagram shown in Figure 2G-1.1. Figure 2G-1.2 shows a block for the telegraph key. A telegraph key is not usually shown as a block. A block generally indicates a stage. A telegraph key is simply a part, similar to an on-off switch. The question probably wants to emphasize the fact that this is a CW transmitter.

Question 2G-1.3. Draw a block diagram representing the stages in a simple telegraphy transmitter having a variable frequency oscillator. /9, 10.

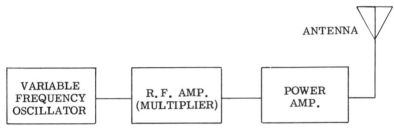

Fig. 2G-1.3. A variable frequency transmitter.

See Figure 2G-1.3. This transmitter is similar in all respects to the transmitter of Figure 2G-1.1, except that a variable frequency oscillator has replaced the crystal controlled oscillator. The variable frequency oscillator allows the operator to change his frequency by means of a knob instead of having to change crystals. This allows for easier operating. While the variable frequency oscillator is not as stable as a crystal controlled oscillator, it is quite satisfactory.

Question 2G-1.4. What type of transmitter does this block diagram (Figure 2G-1.4) represent?

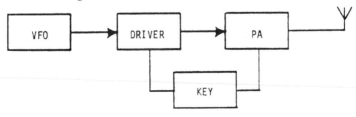

Fig. 2G-1.4.

A. A single conversion receiver
B. A variable frequency oscillator
C. A crystal-controlled transmitter
D. A simple transmitter having a variable frequency oscillator

The answer is D. This is a block diagram of a simple CW transmitter, using a variable frequency oscillator. It is the same as Figure 2G-1.3, except that a block is shown for the telegraph key. See answers to Questions 2G-1.2 and 2G-1.3.

The driver stage is actually an RF amplifier. It builds up the signal and feeds ("drives") the power amplifier.

Question 2G-2.1. Draw a block diagram representing the stages in a simple superheterodyne receiver capable of receiving A1A telegraphy signals. 12/13.

See Figure 2G-2.1.

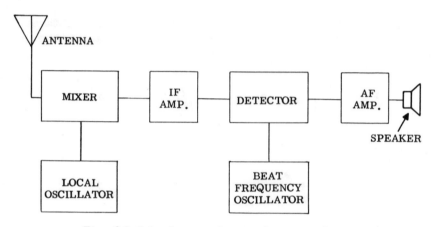

Fig. 2G-2.1. A superheterodyne receiver.

The incoming signal strikes the antenna and is fed to the mixer. At the mixer, the signal is mixed or "heterodyned" with a signal generated by the local oscillator. This produces two new frequencies, the sum and the difference of the incoming signal and the local oscillator frequencies. It is the difference or "intermediate frequency" (IF) in which we are interested. The intermediate frequency is amplified in the IF amplifier stage. The signal is then fed to the detector where it is detected (demodulated). In the detection process, the desired audio is extracted from the high frequency carrier signal. The detected signal or audio is amplified in the audio frequency (AF) amplifier stage. The output of the AF amplifier is fed to a speaker or headset which changes the audio signal into sound.

A CW signal (A1A emission) does not have any audio. The purpose of the beat frequency oscillator (BFO) is to generate a signal which will beat with the incoming RF signal and provide us with a new "difference" frequency which is in the audio range. In this manner, the receiver can detect CW signals. In order to detect AM signals, the beat frequency oscillator is not necessary and is disconnected.

Question 2G-2.2. What type of device does this block diagram (Figure

2G-2.2) represent?

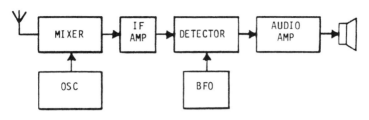

Fig. 2G-2.2.

A. A double conversion receiver
B. A variable frequency oscillator
C. A simple superheterodyne receiver
D. A simple cw transmitter
   The answer is C. Figure 2G-2.2 is identical to Figure 2G-2.1, and the explanation in answer 2G-2.1 applies to Figure 2G-2.2.

Question 2G-3.1. Draw a block diagram representing how two different antennas and a dummy load can be connected to the same transceiver.
   See Figure 2G-3.1. The antenna switch allows the operator to switch the transceiver output between either one of the two antennas or the dummy load. In actual practice, we can tune up the transceiver using the dummy load; then we switch over to the desired antenna. In Figure 2G-3.1, the transceiver is hooked up to antenna #1.

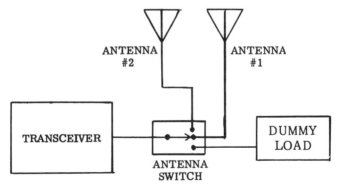

Fig. 2G-3.1. A transceiver with 2 antennas
and a dummy load

Question 2G-3.2. What is the unlabeled (?) block in this diagram (Figure 2G-3.2)?
A. Pi network   B. Antenna switch
C. Key click filter   D. Mixer
   The answer is B. This diagram is similar to the diagram of Figure 2G-3.1. It illustrates how a transceiver can be switched to either

one of two antennas or a dummy load. The unlabeled block is the antenna switch. It is a simple three-way switch that allows the transceiver to be connected to one of the three blocks. The block labeled "dipole" is a simple type of half-wave dipole antenna. The block labeled "beam" is a more complex antenna that concentrates the RF energy radiated from the antenna into a narrow "beam" of energy. The box labeled "dummy" is the dummy load that absorbs the RF output of the transmitter without radiating it. The dummy load is used when tuning up the transmitter.

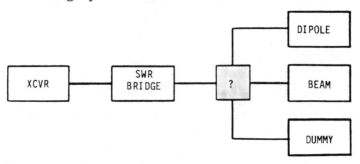

Fig. 2G-3.2

Question 2G-3.3. Draw a block diagram representing an amateur station including transmitter, receiver, telegraph key, TR switch, standing wave ratio bridge, antenna tuner and antenna?

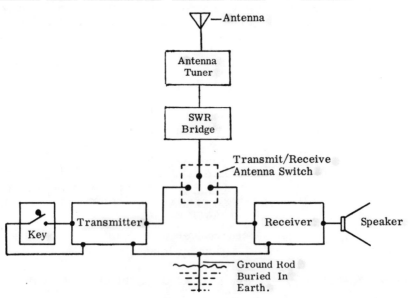

Fig. 2G-3.3 An Amateur Radio Station.

See Figure 2G-3.3. Figure 2G-3.3 illustrates a block diagram of a complete Amateur Radio Station. Note that the transmitter, receiver and telegraph key are all grounded. This will result in better safety and less harmonic radiation.

Question 2G-3.4. What is the unlabeled (?) block in this diagram (Figure 2G-3.4)?

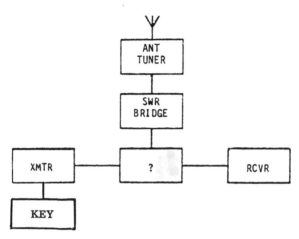

Fig. 2G-3.4.

A. TR switch   B. Variable frequency oscillator
C. Linear amplifier   D. Microphone
   The answer is A. Figure 2G-3.4 is similar to Figure 2G-3.3. It is a block diagram of a complete radio station. The unlabeled block is the Transmit-Receive antenna switch. It switches the antenna, tuner and SWR bridge from the receiver to the transmitter and back.

One (1) question must be from the following:

Question 2G-4.1. In an Amateur Radio station designed for radiotelephone operation, what station accessory will you need to go with your transmitter?
A. A splatter filter   B. A terminal voice controller
C. A microphone   D. A receiver audio filter
   The answer is C. The microphone is the device that transforms sound waves into electrical signals. These audio electrical signals are then fed to an audio amplifier in the transmitter.

Question 2G-4.2. What is the unlabeled block (?) in this block diagram (Figure 2G-4.2) of a radiotelephone station?
A. A splatter filter   B. A terminal voice controller
C. A microphone   D. A receiver audio filter
   The answer is C. See answer 2G-4.1.

Question 2G-5.1. In an Amateur Radio station designed for radiotele-type operation, what station accessories will you need to go with

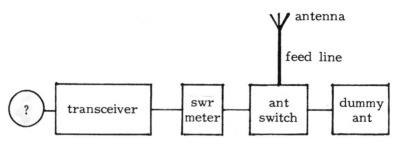

Fig. 2G-4.2

your transmitter?
A. A computer, a printer and a RTTY refresh unit
B. A modem and a teleprinter or computer system
C. A terminal-node controller
D. A modem, a monitor and a DTMF key pad

The answer is B. Radioteletype (abbreviated RTTY) is a system in which one types out information with a device that is similar to a typewriter. This typewriter device changes the letters of the alphabet into electrical signals. These electrical signals are used to modulate a carrier, which is then transmitted. At the receiving end, a receiver picks up and amplifies the signal. The signal is then fed to a demodulator, which extracts the information and changes it into a form which activates the keys of a typewriter and prints out the original information.

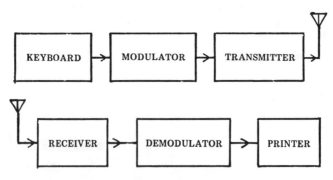

Fig. 2G-5.1. A basic radioteletype system

Figure 2G-5.1 illustrates a simplified block diagram of a complete teletype system. The keyboard is the typewriter device that generates combinations of pulses. Each combination represents a different letter of the alphabet. The modulator converts the pulses into different frequency tones. These tones then modulate the transmitter.

At the receiving end, the receiver changes the radio signals into tones which are then applied to the demodulator. The demodulator is also called a detector, a converter, or a terminal unit. The demodulator converts the tones into corresponding pulses. The pulses are

applied to the printer, which responds by printing a different letter
of the alphabet for each different combination of pulses.

In a modern system, the transmitter and receiver can be housed in
one unit, called a transceiver (XCVR). The modulator and demodulator
can be housed in one unit called a modem (modem is the contraction
of the terms modulator and demodulator). The keyboard and printer
are part of the teleprinter or computer system. See Figure 2G-5.2.

**Question 2G-5.2. Draw a block diagram showing how the parts of a
radioteletype station connect. Include at least a modem, transceiver,
computer system or teleprinter, feed line and antenna.**
See Figure 2G-5.2 and answer 2G-5.1.

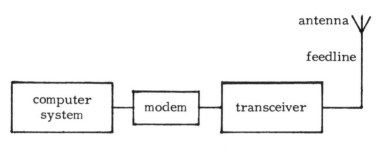

Fig. 2G-5.2.

**Question 2G-6.1. In a packet-radio station, what device connects
between the radio transceiver and the computer terminal?**
A. An RS-232 interface
B. A terminal-node controller
C. A terminal refresh unit
D. A tactical network control system
The answer is B. See answer 2B-9.1.

**Question 2G-6.2. What is the unlabeled block (?) in this diagram
(Figure 2G-6.2) of a packet-radio station?**

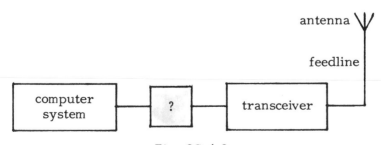

Fig. 2G-6.2.

A. An RS-232 interface
B. A terminal-node controller

C. A terminal refresh unit
D. A tactical network control system
    The answer is B. See answer 2B-9.1.

Question 2G-6.3. Where does a terminal-node controller connect in an amateur packet-radio station?
A. Between the antenna and the radio
B. Between the computer and the monitor
C. Between the computer or terminal and the radio
D. Between the keyboard and the computer
    The answer is C. See answers 2G-6.2 and 2B-9.1.

SUBELEMENT 2H
Signals and Emissions
(2 questions)

One (1) question must be from the following:

Question 2H-1.1. Which type of emission is an interrupted carrier wave? 9,10/10,11.
A. A1A   B. A3J   C. F3C   D. F2B
The answer is A. We also refer to this type of emission as Continuous Wave (CW).

Question 2H-2.1. What does the term backwave mean? 9/10.
A. A radio wave reflected from the ionosphere back to the sending station
B. A small amount of RF that a CW transmitter produces even when the key is not closed
C. Radio waves reflected back down the feed line from a mismatched antenna
D. The reflected power in a feed line
The answer is B. "Backwave" is the radiation of RF energy from the antenna when the key is up. There should be NO radiation of RF energy when the key is up. A signal with backwave is difficult to read and should be avoided.

Question 2H-2.2. What is a possible cause of backwave? 9/10.
A. Low voltage   B. Poor neutralization
C. Excessive RF drive   D. Mismatched antenna
The answer is B. Other causes of backwave are keying of the final amplifier stage only, and inductive pickup between the antenna coupling coils and one of the lower power stages.

Question 2H-3.1. What does the term key click mean? 9/10.
A. The mechanical noise caused by a straight key
B. An excessively square CW keyed waveform
C. An excessively fast CW signal
D. The sound of a CW signal being copied on an AM receiver
The answer is B. Key clicks are the loud "clicks" or "thumps" that occur at the beginning and/or end of each dot and dash. They are due to the fact that large amounts of power are suddenly applied and removed from a circuit when a transmitter is keyed. See the answer to Question 2H-3.2.

Question 2H-3.2. How can key clicks be eliminated? 9/10.
A. By carefully adjusting your antenna matching network
B. By increasing power to the maximum allowable level
C. By using a power supply with better regulation

D. By using a key-click filter

The answer is D. Figure 2H-3.2 illustrates the circuit of a key-click filter. Inductance, L, causes a slight lag in current and prevents the sudden current build-up when the key is closed. The combination of C and R absorb the sparks that occur across the key when it is opened and closed. Sparks cause the generation of sidebands which produce clicks in receivers, even though they are not tuned to the same frequency as the offending transmitter.

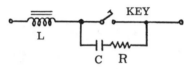

Fig. 2H-3.2 A key-click filter.

Question 2H-4.1. What does the term chirp mean? 9/10.
A. A distortion in the receiver audio circuits
B. A high-pitched audio tone transmitted with a CW signal
C. A slight shift in oscillator frequency each time a CW transmitter is keyed
D. A slow change in transmitter frequency as the circuit warms up

The answer is C. A "chirp" occurs when the tone of the dot or dash changes. The changing tone sounds like a chirp. Chirping is due to a changing oscillator frequency when the key is depressed. A changing oscillator frequency is due to a change in voltage on the elements of the oscillator's tube or transistor.

Question 2H-4.2. What can be done to a telegraph transmitter power supply to avoid chirp? 9/10.
A. Resonate the power supply filters
B. Regulate the power supply output voltages
C. Use a buffer amplifier between the transmitter output and the feed line
D. Cause the power-supply output current to vary with the load

The answer is B. The regulation of the power supply (its ability to maintain a constant output voltage, regardless of load current changes) should be improved. This will cause the output voltage of the power supply, which is applied to the oscillator stage, to remain constant when the key is depressed and thereby eliminate chirp.

Question 2H-5.1. What is a common cause of superimposed hum? 9/10.
A. Using a nonresonant random-wire antenna
B. Sympathetic vibrations from a nearby transmitter
C. Improper neutralization of the transmitter output stage
D. A defective filter capacitor in the power supply

The answer is D. Replacing the defective capacitor will remove the hum from the signal.

Question 2H-6.1. 28.160 MHz is the 4th harmonic of what fundamental frequency? 9/8,10.
A. 7.040 MHz   B. 112.64 MHz   C. 7.160 MHz   D. 1.760 MHz

The answer is A. Since we multiply by 4 to get the fourth harmonic of a fundamental signal, we simply divide by 4 to get the fundamental that causes the fourth harmonic.

Question 2H-7.1. What problem in a transmitter power amplifier stage may cause spurious emissions? /10.
A. Excessively fast keying speed
B. Undermodulation
C. Improper neutralization
D. Tank circuit current dip at resonance

The answer is C. Spurious emissions may also be caused by improper parts placement, improper lead dress, or lack of proper shielding and grounding in the power amplifier stage.

Question 2H-8.1. What emission designator describes the use of frequency shift keying to transmit radioteletype messages?
A. F2D   B. F1B   C. J1F   D. A1B

The answer is B. See chart on page AP-1.

Question 2H-8.2. What keying method is used to transmit F1B radioteletype messages?
A. Frequency shift keying
B. On-off keying of the radio wave
C. Split-baud keying
D. Tuned-output keying

The answer is A. In frequency shift keying (abbreviated FSK), we take a CW signal that is "on" all the time and vary it between two frequencies. The difference between the two frequencies (frequency shifting) is usually 170 Hz. This frequency shift must be absolutely constant so that the selective filters in the receiving converter can respond properly. Frequency shift keying is used below 50 MHz (HF bands).

Question 2H-9.1. What emission designator describes single-sideband suppressed-carrier (SSB) voice transmissions?
A. J2D   B. A3J   C. J3E   D. F3E

The answer is C. See chart on page AP-1.

Question 2H-9.2. What type of signal is emission J3E?
A. Frequency-modulated voice
B. Single-sideband suppressed-carrier voice
C. Frequency-shift keyed RTTY
D. Packet radio

The answer is B. Single sideband suppressed carrier voice signals occupy half the bandwidth that AM signals occupy. Also, SSB signals require far less power than AM signals for transmitting a certain amount of information.

Question 2H-10.1. What emission designator describes frequency-modulated voice transmissions?
A. F1B   B. F2D   C. F3E   D. A3F
   The answer is C. See chart on page AP-1.

Question 2H-10.2. What type of signal is emission F3E?
A. Frequency-modulated voice
B. Single-sideband suppressed-carrier voice
C. Frequency-shift keyed RTTY
D. Packet radio
   The answer is A. In a frequency modulated system, the carrier amplitude remains constant and its frequency varies. This is in contrast with an AM system where the carrier frequency is constant and the carrier's amplitude varies.

One (1) question must be from the following:

Question 2H-11.1. What may happen to body tissues that are exposed to large amounts of RF energy?
A. The tissue may be damaged because of the heat produced
B. The tissue may suddenly be frozen
C. The tissue may be immediately destroyed because of the Maxwell effect
D. The tissue may become less resistant to cosmic radiation
   The answer is A. Exposure to large amounts of RF, especially VHF and UHF will cause tissue damage. See answer 2D-9.2.

Question 2H-11.2. What precaution should you take before working near a high-gain UHF or microwave antenna (such as a parabolic, or dish antenna)?
A. Be certain the antenna is FCC type approved
B. Be certain the antenna and transmitter are properly grounded
C. Be certain the transmitter cannot be operated
D. Be certain the antenna safety interlocks are in place
   The answer is C. See answer 2D-9.2.

Question 2H-11.3. How should the antenna on a hand-held transceiver be positioned while you are transmitting?
A. As close to your body as possible, to take advantage of the directional effect
B. Away from your head and away from others standing nearby, to minimize RF exposure
C. Close to the ground, since a hand-held transceiver has no ground connection
D. As close to a vertical position as possible, to minimize corona effect
   The answer is B. See answer 2D-9.2.

Question 2H-11.4. Why should you always locate your antennas so that no one can come in contact with them while you are transmitting?
A. To prevent damage to the antennas
B. To prevent RF burns and excessive exposure to RF energy
C. To comply with FCC regulations concerning antenna height
D. To prevent unexpected changes in your standing-wave ratio
   The answer is B. See answer 2D-9.2.

Question 2H-11.5. What is a good way to prevent RF burns and excessive exposure to RF from your antennas?
A. Shield your antenna with a grounded RF screen
B. Make sure you use plenty of radial wires in your antenna installation
C. Use burn-proof wire for your antenna feed line
D. Always locate your antennas so that no one can come in contact with them while you are transmitting
   The answer is D. See answer 2D-9.2.

Question 2H-12.1. What type of interference will you cause if you operate your SSB transmitter with the microphone gain adjusted too high?
A. You may cause digital interference to computer equipment in your neighborhood
B. You may cause atmospheric interference in the air around your antenna
C. You may cause splatter interference to other stations on nearby frequencies
D. You may cause processor interference to the microprocessor in your rig
   The answer is C. Splatter is a term used to describe severe distortion and adjacent channel interference. Proper adjustment of the microphone gain control prevents this distortion by automatically limiting the gain of prior stages. However, the audio gain should be high enough for proper modulation.

Question 2H-12.2. What may happen if you adjust the microphone gain or deviation control on your FM transmitter too high?
A. You may cause digital interference to computer equipment in your neighborhood
B. You may cause interference to other stations on nearby frequencies
C. You may cause atmospheric interference in the air around your antenna
D. You may cause processor interference to the microprocessor in your rig
   The answer is B. By limiting the audio gain, we prevent excessive deviation with its subsequent interference to adjacent frequencies. On the other hand, the· audio should be strong enough to provide reasonable deviation which produces a strong FM signal.

Question 2H-12.3. If you are using an excessive amount of speech processing with your SSB transmitter, what type of interference are you likely to cause?
A. You may cause digital interference to computer equipment in your neighborhood
B. You may cause splatter interference to other stations on nearby frequencies
C. You may cause atmospheric interference in the air around your antenna
D. You may cause processor interference to the microprocessor in your rig
The answer is B. See answer 2H-12.1.

Question 2H-12.4. If you are operating SSB voice and another operator tells you that you are causing "splatter", what might be the cause of the interference?
A. Your rig may be switching from transmit to receive too quickly
B. You may have your transmitter microphone gain control set too high
C. Your rig may have a defective modulator transistor
D. You may have your transmitter splatter control set incorrectly
The answer is B. See answer 2H-12.1.

Question 2H-12.5. If you are operating FM voice and another operator tells you that your signal is "too wide" and that you are causing interference to other stations on nearby frequencies, what might be the cause of the interference?
A. You may have your transmitter deviation control or microphone gain control set too high
B. The spectral width control on your transmitter may be set incorrectly
C. Your microphone may be defective
D. You may need to use an amplified "power microphone"
The answer is A. See answer 2H-12.2.

One (1) question must be from the following:

Question 2I-1.1. What is the approximate length of a half-wave dipole antenna for 3725-kHz? 11/12.
A. 126 ft   B. 81 ft   C. 63 ft   D. 40 ft
    The answer is A. A reasonably accurate formula for finding the length of a half-wave antenna where the frequency is known, is:

$$L = \frac{468}{f}$$ where: L is the antenna length in feet, and f is the frequency in MHz. If the frequency is given in kiloHertz (kHz), we must change it to MegaHertz (MHz) before substituting in the formula.

    We solve the above problem as follows: We change kHz to MHz by dividing the kHz by 1,000.

$$\frac{3,725 \text{ kHz}}{1,000} = 3.725 \text{ MHz}$$

We then substitute 3.725 for f in the formula:

$$L = \frac{468}{3.725} = 125.64 \text{ feet}$$

Question 2I-1.2. What is the approximate length of a half-wave dipole antenna for 7125-kHz? 11/12.
A. 84 ft   B. 42 ft   C. 33 ft   D. 66 ft
    The answer is D. We find the antenna length by using the method given in answer 2I-1.1.

$$\frac{7,125 \text{ kHz}}{1,000} = 7.125 \text{ MHz} \qquad L = \frac{468}{7.125} = 65.68 \text{ feet}$$

Question 2I-1.3. What is the approximate length of a half-wave dipole antenna for 21,125-kHz? 11/12.
A. 44 ft   B. 28 ft   C. 22 ft   D. 14 ft
    The answer is C. We find the antenna length using the method given in Question 2I-1.1.

$$\frac{21,125 \text{ kHz}}{1,000} = 21.125 \text{ MHz} \qquad L = \frac{468}{21.125} = 22.15 \text{ feet}$$

Question 2I-1.4. What is the approximate length of a half-wave dipole antenna for 28,150-kHz? 11/12.
A. 22 ft   B. 11 ft   C. 17 ft   D. 34 ft
    The answer is C. We find the antenna length using the method given in the answer to Question 2I-1.1.

$$\frac{28,150 \text{ kHz}}{1,000} = 28.150 \qquad L = \frac{468}{28.15} = 16.63 \text{ feet}$$

Question 2I-1.5. How is the approximate length of a half-wave dipole antenna calculated? 11/12.

A. By substituting the desired operating frequency for f in the formula: 150 / f (MHz)
B. By substituting the desired operating frequency for f in the formula: 234 / f (MHz)
C. By substituting the desired operating frequency for f in the formula: 300 / f (MHz)
D. By substituting the desired operating frequency for f in the formula: 468 / f (MHz)

The answer is D. See answer to question 2I-1.1.

Question 2I-2.1. What is the approximate length of a quarter-wave vertical antenna adjusted to resonate at 3725-kHz? 11/12.

A. 20 ft   B. 32 ft   C. 40 ft   D. 63 ft

The answer is D. This question is similar to question 2I-1.1, except that we are looking for the length of a quarter-wave instead of a half-wave antenna. Since a quarter-wave antenna is one half of a half-wave antenna, we can use the same formula, and then divide the answer in half.

The length of a quarter-wave antenna for 3,725 kHz would, therefore, be half of that given in the answer to question 2I-1.1, or 125.64 divided by 2 = 62.82 feet.

We could also find the length of a quarter-wave antenna by cutting the numerator of the formula in half. Thus, the formula for the length of a quarter-wave antenna would be:

$$L = \frac{234}{f}$$ where: L is the antenna length in feet and f is the frequency in MHz.

If we use this formula, we will come up with the same answer as above:

$$L = \frac{234}{3.725} = 62.82 \text{ feet}$$

Question 2I-2.2. What is the approximate length of a quarter-wave vertical antenna adjusted to resonate at 7125-kHz? 11/12.

A. 11 ft   B. 16 ft   C. 21 ft   D. 33 ft

The answer is D. It would be one half of the answer to question 2I-1.2, or 65.68 feet divided by 2 = 32.84 feet. See answer 2I-2.1.

Question 2I-2.3. What is the approximate length in feet of a quarter-wave vertical antenna adjusted to resonate at 21,125-kHz? 11/12.

A. 7 ft   B. 11 ft   C. 14 ft   D. 22 ft

The answer is B. It would be one half of the answer to question 2I-1.3, or 22.15 feet divided by 2 = 11.08 feet. See answer 2I-2.1.

Question 2I-2.4. What is the approximate length of a quarter-wave vertical antenna adjusted to resonate at 28,150-kHz? 11/12.
A. 5 ft   B. 8 ft   C. 11 ft   D. 16 ft
The answer is B. It would be one half of the answer to question 2I-1.4, or 16.63 feet divided by 2 = 8.32 feet. See answer 2I-2.1.

Question 2I-2.5. When a vertical antenna is lengthened, what happens to its resonant frequency?   /12.
A. It decreases   B. It increases
B. It stays the same   D. It doubles
The answer is A. The resonant frequency of an antenna varies inversely with its length. Thus, as the antenna is lengthened, its resonant frequency is reduced. If the antenna length is reduced, its resonant frequency is increased.

One (1) question must be from the following:

Question 2I-2.6. What is the approximate length (in inches) of a 5/8-wavelength vertical antenna for the 220-MHz band?
A. 19-1/2 inches   B. 22 inches
C. 28-1/2 inches   D. 32 inches
The answer is C. If we use the half wavelength formula of answer 2I-1.1 and adjust the answer for a 5/8-wavelength antenna, we arrive at 32 inches. However, when feeding a 5/8-wavelength antenna, we must use an impedance matching coil which electrically "lengthens" the antenna. To compensate, we shorten the physical length of the antenna, in this case, to 28-1/2 inches.

Question 2I-2.7. Why do many amateurs use a 5/8-wavelength vertical antenna rather than a 1/4-wavelength vertical antenna for their VHF or UHF mobile stations?
A. A 5/8-wavelength antenna can handle more power than a 1/4-wavelength antenna
B. A 5/8-wavelength antenna has more gain than a 1/4-wavelength antenna
C. A 5/8-wavelength antenna exhibits less corona loss than a 1/4-wavelength antenna
D. A 5/8-wavelength antenna looks more like a CB antenna, so it does not attract as much attention as a 1/4-wavelength antenna
The answer is B. A 5/8-wavelength vertical antenna has a power gain of 3 dB over a 1/4 wavelength vertical antenna.

Question 2I-3.1. What is a <u>coaxial</u> cable? 11/12.
A. Two parallel conductors encased along the edges of a flat plastic ribbon
B. Two parallel conductors held at a fixed distance from each other by insulating rods
C. Two conductors twisted around each other in a double spiral
D. A center conductor encased in insulating material which is covered by a conducting sleeve or shield, and encased in a weatherproof

jacket
The answer is D. Coaxial cable is a commonly used transmission line (feedline) that conducts RF energy from the transmitter to its antenna. It also conducts a signal from the antenna to its receiver. Coaxial cable consists of an inner conductor surrounded by a flexible insulator, which, in turn, is surrounded by a flexible wire braid. A weatherproof vinylite sheath surrounds the flexible wire braid. See Figure 2I-3.1. Typical characteristic impedances of coaxial cable are 52 and 75 ohms.

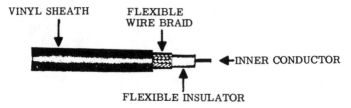

VINYL SHEATH   FLEXIBLE
                      WIRE BRAID

◄—INNER CONDUCTOR

FLEXIBLE INSULATOR

Fig. 2I-3.1 Coaxial Cable.

Question 2I-3.2. What kind of antenna feedline is constructed of a center conductor encased in insulation which is then covered by an outer conducting shield and weatherproof jacket? 11/12.
A. Twin lead   B. Coaxial cable
C. Open-wire feed line   D. Waveguide
     The answer is B. See the answer to question 2I-3.1.

Question 2I-3.3. What are some advantages in using coaxial cable as an antenna feedline?   /12.
A. It is easy to make at home, and it has a characteristic impedance in the range of most common amateur antennas
B. It is weatherproof, and it has a characteristic impedance in the range of most common amateur antennas
C. It can be operated at a higher SWR than twin lead, and it is weatherproof
D. It is unaffected by nearby metallic objects, and has a characteristic impedance that is higher than twin lead
     The answer is B. Other advantages of using coaxial cable are:
     (1) The outer wire braid acts as a shield, preventing the radiation of harmonic and spurious signals.
     (2) It is simple to install.
     (3) It has minimum noise pickup.
     (4) It is efficient.

Question 2I-3.4. What commonly-available antenna feedline can be buried directly in the ground for some distance without adverse effects?
A. Twin lead   B. Coaxial cable
C. Parallel conductor   D. Twisted pair
     The answer is B. Coaxial cable can be used because of its

weatherproof vinyl covering.

Question 2I-3.5. When an antenna feedline must be located near grounded metal objects, which commonly-available feedline should be used?
A. Twisted pair   B. Twin lead
C. Coaxial cable   D. Ladder-line
    The answer is C. Coaxial cable should be used because its outer conductor acts as a shield.

One (1) question must be from the following:

Question 2I-4.1. What is parallel conductor feedline? 11/12.
A. Two conductors twisted around each other in a double spiral
B. Two parallel conductors held a uniform distance apart by insulating material
C. A conductor encased in insulating material which is then covered by a conducting shield and a weatherproof jacket
D. A metallic pipe whose diameter is equal to or slightly greater than the wavelength of the signal being carried
    The answer is B. There are two types of parallel conductor feedline. One is called the "open wire" type, and the other is called the "twin-lead" type. Open-wire transmission line consists of two conductors in parallel, separated by insulating spacers. The twin-lead type consists of two parallel conductors, separated by flexible insulation. This type is commonly used in television installations. See Figure 2I-4.1.

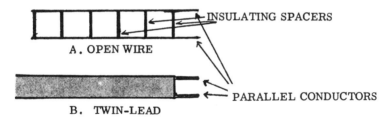

Fig. 2I-4.1 Parallel conductor transmission line.

Question 2I-4.2. How can TV-type twin lead be used as a feedline? 11/
A. By carefully running the feed line parallel to a metal post to ensure self resonance
B. TV-type twin lead cannot be used in an Amateur Radio station
C. By installing an impedance-matching network between the transmitter and feed line
D. By using a high-power amplifier and installing a power attenuator between the transmitter and feed line
    The answer is C. TV-type twin lead can be used in the same manner as the better quality amateur parallel twin lead is used.

However, the amateur must take into account the impedance of the TV-type twin lead. Most TV-type twin lead is 300 ohms; some of it is 75 ohms. He must make certain that the impedance of the transmitter, the twin lead and the antenna are properly matched. An impedance matching network may be necessary.

There are many brands of TV-type twin lead. An amateur should use a good quality standard brand so that the feedline losses are low. He should also check the current rating of the twin lead to make certain that it can handle the current that will be drawn by the antenna.

Question 2I-4.3. What are some advantages in using a parallel conductor feedline? /12.
A. It has a lower characteristic impedance than coaxial cable, and will operate at a higher SWR than coaxial cable
B. It will operate at a higher SWR than coaxial cable, and it is unaffected by nearby metal objects
C. It has a lower characteristic impedance than coaxial cable, and has less loss than coaxial cable
D. It will operate a at higher SWR than coaxial cable and it has less loss than coaxial cable

The answer is D. Some other advantages are:

(1) It is a balanced line and is easier to match to a dipole antenna which is also balanced.

(2) The open wire type has a characteristic impedance of from 200 to 600 ohms, and can be used to match high impedance antennas.

(3) It has very low losses if it is properly constructed and properly used.

Question 2I-4.4. What are some disadvantages in using a parallel conductor feedline? /12
A. It is affected by nearby metallic objects, and it has a characteristic impedance that is too high for direct connection to most amateur transmitters
B. It is more difficult to make at home than coaxial cable and it cannot be operated at a high SWR
C. It is affected by nearby metallic objects, and it cannot handle the power output of a typical amateur transmitter
D. It has a characteristic impedance that is too high for direct connection to most amateur transmitters, and it will operate at a high SWR

The answer is A. Some disadvantages are:

(1) It is not shielded and can radiate or be affected by nearby metal objects.

(2) It may be difficult to install because the wires get tangled and touch each other.

(3) The twin lead type exhibits very high losses in wet weather.

Question 2I-4.5. What kind of antenna feedline is constructed of two conductors maintained a uniform distance apart by insulated spreaders? 11/12.
A. Coaxial cable   B. Ladder-line open-conductor line
C. Twin lead in a plastic ribbon D. Twisted pair
    The answer is B. It is called "open-wire, parallel", or simply "open-wire" transmission line.

Question 2I-5.1. What type of radiation pattern is produced by a 5/8-wavelength vertical antenna?
A. A pattern with the transmitted signal spread out equally in all directions
B. A pattern with more of the transmitted signal concentrated in one direction than in other directions.
C. A pattern with most of the transmitted signal concentrated in two opposite directions
D. A pattern with most of the transmitted signal concentrated at high radiation angles
    The answer is A. A vertical antenna radiates equally well in all horizontal directions.

Question 2I-6.1. What type of radiation pattern is produced by a Yagi antenna?
A. A pattern with the transmitted signal spread out equally in all directions
B. A pattern with more of the transmitted signal concentrated in one direction than in other directions.
C. A pattern with most of the transmitted signal concentrated in two opposite directions
D. A pattern with most of the transmitted signal concentrated at high radiation angles
    The answer is B. A Yagi antenna consists of an ordinary half-wave dipole, plus one or more elements. Figure 2I-6 illustrates a simple three element Yagi antenna. The half-wave dipole element (B in Figure 2I-6), to which the transmission line is connected, is called the DRIVEN element. The driven element receives the transmitter output power from the transmission line. The other elements (A and C) are not physically connected to the transmission line or to the driven element. They are called "parasitic elements", and they receive their energy from the electromagnetic radiation of the driven element.
    The addition of the parasitic elements increases the gain of the antenna considerably over that of a simple single element dipole. The additional antenna elements don't add RF energy to the antenna; they concentrate the radiated energy in one direction. The more the parasitic elements, the greater is the concentration or directivity of the antenna.
    The transmitting advantages of a Yagi are also present in receiving. It will receive much better in the direction that it is

"pointed", compared to a single element dipole. Also, it does not pick up well from other directions.

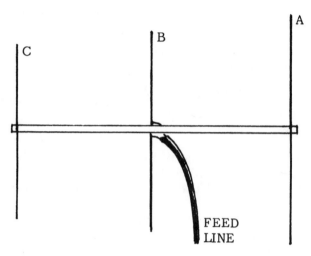

Fig. 2I-6.

Question 2I-6.2. On the Yagi antenna shown in Figure 2I-6, what is the name of section B?
A. Director   B. Reflector   C. Boom   D. Driven element
    The answer is D. See answer 2I-6.1.

Question 2I-6.3. On the Yagi antenna shown in Figure 2I-6, what is the name of section C?
A. Director   B. Reflector   C. Boom   D. Driven element
    The answer is A. The director is about 5% shorter than the driven element. The director is in the direction that most of the transmitted energy will travel.

Question 2I-6.4. On the Yagi antenna shown in Figure 2I-6, what is the name of section A?
A. Director   B. Reflector   C. Boom   D. Driven element
    The answer is B. The reflector is about 5% longer than the driven element.

Question 2I-6.5. Approximately how long (in wavelengths) is the driven element of a Yagi antenna?
A. 1/4 wavelength   B. 1/3 wavelength
C. 1/2 wavelength   D. 1 wavelength
    The answer is C. The driven element, by itself, is a half-wave dipole.

SAMPLE NOVICE EXAMINATION
(Answers on page SE-5)

The following sample examination provides an added means of preparation for the actual test and serves as a guage of preparadness.

1. What is <u>amateur</u> radiocommunication?
A. Non-commercial radio communication between Amateur Radio stations with a personal aim and without pecuniary interest
B. Commercial radio communications between radio stations licensed to non-profit organizations and businesses
C. Experimental or educational radio transmissions controlled by student operators
D. Non-commercial radio communications intended for the education and benefit of the general public

2. What are the Novice control operator frequency privileges in the 40 meter band?
A. 3500 to 4000 kHz   B. 3700 to 3750 kHz
C. 7100 to 7150 kHz   D. 7000 to 7300 kHz

3. What does the term <u>CW</u> mean?
A. Calling wavelength   B. Coulombs per watt
C. Continuous wave   D. Continuous wattage

4. On what frequencies in the 10-meter band are Novice control operators permitted to transmit emission F1B (RTTY)?
A. 28.1 to 28.5 MHz   B. 28.0 to 29.7 MHz
C. 28.1 to 28.2 MHz   D. 28.1 to 28.3 MHz

5. What is the meaning of the term <u>malicious interference?</u>
A. Accidental interference   B. Intentional interference
C. Mild interference   D. Occasional interference

6. With which non-amateur radio stations may an FCC-licensed amateur station communicate?
A. No non-amateur stations
B. All such stations
C. Only those authorized by the FCC
D. Only those who use the International Morse code

7. What station identification, if any, is required at the beginning of a QSO?
A. The operator originating the contact must transmit both call signs
B. No identification is required at the beginning of the contact
C. Both operators must transmit their own call signs
D. Both operators must transmit both call signs

8. What is the amount of transmitting power that an amateur station must never exceed when transmitting on 3725-kHz?

A. 75 watts PEP output   B. 100 watts PEP output
C. 200 watts PEP output   D. 1500 watts PEP output

9. What term is used to describe amateur communications for the direct transfer of information between computers?
A. Teleport communications   B. Direct communications
C. Digital communications   D. Third-party communications

10. What is the meaning of the term zero beat?
A. Transmission and reception on the same operating frequency
B. Transmission on a predetermined frequency
C. Used only for satellite reception
D. Unimportant for CW operations

11. What is the commonly used RTTY sending speed above 50 MHz?
A. 1200 bauds   B. 60 bauds
D. 100 bauds   D. 9600 bauds

12. What does the term skip mean?
A. Signals are reflected from the moon
B. Signals are refracted by water-dense cloud formations
C. Signals are retransmitted by repeaters
D. Signals are refracted by the ionosphere

13. What type of antenna polarization is normally used for communications on the 80-meter band?
A. Right-hand circular polarization
B. Magnetic polarization
C. Horizontal polarization
D. Vertical polarization

14. For proper protection from lightning strikes, what pieces of equipment should be grounded in an amateur station?
A. The power supply primary
B. All station equipment
C. The feed line center conductors
D. The ac power mains

15. What type of filter should be installed on a TV receiver tuner as the first step in preventing overload from an amateur station transmission?
A. Low pass   B. High pass   C. Band pass   D. Notch

16. If the standing wave ratio bridge reading is higher at 3700-kHz than at 3750-kHz, what does this indicate about the antenna?
A. Too long for optimal operation at 3700 kHz
B. Broadbanded
C. Good only for 37-meter operation
D. Too short for optimal operation at 3700 kHz

17. What precautions should you take before removing the shielding on a VHF or UHF power amplifier?
A. Make sure all RF screens are in place at the antenna
B. Make sure the feed line is properly grounded
C. Make sure the amplifier cannot be accidentally energized
D. Make sure that the RF leakage filters are connected

18. What type of current changes direction over and over again in a cyclical manner?
A. Direct current
B. Alternating current
C. Negative current
D. Positive current

19. What is the approximate length, in meters, of a radio wave having a frequency of 3.725-MHz?
A. 160 meters   B. 80 meters   C. 40 meters   D. 30 meters

20. What type of frequency is 350 Hertz?
A. An audio frequency
B. A microwave frequency
C. An intermediate frequency
D. A radio frequency

21. What do the prefixes mega- and centi- mean?
A. 1,000,000 and 0.01
B. 0.001 and 0.01
C. 1,000,000 and 100
D. 0.001 and 100

22. What is the general relationship between the thickness of a quartz crystal and its fundamental operating frequency?
A. The thickness of a crystal does not affect operating frequency
B. Thinner crystals oscillate at lower frequencies
C. Thinner crystals oscillate at higher frequencies
D. Thicker crystals oscillate at higher frequencies

23. What device should be included in electronic equipment to protect it from damage resulting from a short circuit?
A. Fuse   B. Tube   C. Transformer   D. Filter

24. What type of device does this block diagram (Figure 24) represent?

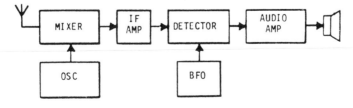

Fig. 24

A. A double conversion receiver

B. A variable frequency oscillator
C. A simple superheterodyne receiver
D. A simple cw transmitter

25. In an Amateur Radio station designed for radioteletype operation, what station accessories will you need to go with your transmitter?
A. A computer, a printer and a RTTY refresh unit
B. A modem and a teleprinter or computer system
C. A terminal-node controller
D. A modem, a monitor and a DTMF key pad

26. What is a possible cause of backwave?
A. Low voltage
B. Poor neutralization
C. Excessive RF drive
D. Mismatched antenna

27. What is a good way to prevent RF burns and excessive exposure to RF from your antennas?
A. Shield your antenna with a grounded RF screen
B. Make sure you use plenty of radial wires in your antenna installation
C. Use burn-proof wire for your antenna feed line
D. Always locate your antennas so that no one can come in contact with them while you are transmitting

28. What is the approximate length of a half-wave dipole antenna for 21,125-kHz?
A. 44 ft   B. 28 ft   C. 22 ft   D. 14 ft

29. What commonly-available antenna feedline can be buried directly in the ground for some distance without adverse effects?
A. Twin lead
B. Coaxial cable
C. Parallel conductor
D. Twisted pair

30. What type of radiation pattern is produced by a Yagi antenna?
A. A pattern with the transmitted signal spread out equally in all directions
B. A pattern with more of the transmitted signal concentrated in one direction than in other directions.
C. A pattern with most of the transmitted signal concentrated in two opposite directions
D. A pattern with most of the transmitted signal concentrated at high radiation angles

ANSWERS TO SAMPLE NOVICE EXAMINATION

1. A	7. B	13. C	19. B	25. B
2. C	8. C	14. B	20. A	26. B
3. C	9. C	15. B	21. A	27. D
4. D	10. A	16. D	22. C	28. C
5. B	11. A	17. C	23. A	29. B
6. C	12. D	18. B	24. C	30. B

APPENDIX 1
Table of Emissions

The FCC is now using the new WARC emission symbols. In the new system, the old 2-character symbols have been replaced with 3-character symbols. The 3-character symbols give more specific information concerning the emissions that they represent.

FIRST CHARACTER

N   Emission of an unmodulated carrier
A   AM double-sideband
J   Single sideband, suppressed carrier
F   Frequency modulation
P   Sequence of unmodulated pulses
C   Vestigial sidebands

SECOND CHARACTER

0   No modulating symbol
1   Digital information - no modulation
2   Digital information with modulation
3   Modulated with analog information

THIRD CHARACTER

N   No information transmitted
A   Telegraphy for reception by air
B   Telegraphy for automatic reception
C   Facsimile
D   Data transmission, telemetry, telecommand
E   Telephony
F   Television

Traditional Symbol		New Symbol
AMPLITUDE MODULATED		
Unmodulated	A0	NON
Keyed on/off	A1	A1A
Tones keyed on/off	A2	A2A
AM data		A2D
Keyed tones w/SSB	A2J	J2A
SSB data		J2D
AM voice	A3	A3E
Voice w/SSB	A3J	J3E
AM facsimile	A4	A3C
SSB television	A5	C3F
AM television	A5	A3F
FREQUENCY MODULATED		
Unmodulated	F0	NON
Switched between		
two frequencies	F1	F1B
Switched tones	F2	F2A
FM data		F2D
FM voice	F3	F3E
FM facsimile	F4	F3C
FM television	F5	F3F
PULSE MODULATED		
Phase	P	P1B

APPENDIX 2
RST Reporting System

The RST Reporting System is a means of rating the quality of a signal on a numerical basis. In this system, the R stands for readability, and is rated on a scale of 1 to 5. The S stands for signal strength, and it is rated on a scale of 1 to 9. The T indicates the quality of a CW tone, and its scale is also 1 to 9. The higher the number, the better the signal.

READABILITY
1. Unreadable.
2. Barely readable; occasional words distinguishable.
3. Readable with considerable difficulty.
4. Readable with practically no difficulty.
5. Perfectly readable.

SIGNAL STRENGTH
1. Faint; signals barely perceptible.
2. Very weak signals.
3. Weak signals.
4. Fair signals.
5. Fairly good signals.
6. Good signals.
7. Moderately strong signals.
8. Strong signals.
9. Extremely strong signals.

TONE
1. Extremely rough, hissing tone.
2. Very rough AC note; no trace of musicality.
3. Rough, low-pitched AC note; slightly musical.
4. Rather rough AC note; moderately musical.
5. Musically modulated.
6. Modulated note; slight trace of whistle.
7. Near DC note; smooth ripple.
8. Good DC note; just a trace of ripple.
9. Purest DC note. (If note appears to be crystal controlled, add letter X after the number indicating tone).

EXAMPLE; Your signals are RST 599X. (Your signals are perfectly readable, extremely strong, have purest DC note, and sound as if your transmitter is crystal-controlled).

FREQUENCY ALLOCATIONS FOR POPULAR AMATEUR BANDS
All in MegaHertz. "X" indicates no privileges.

CLASSES	NOVICE		TECHNICIAN		GENERAL AND CONDITIONAL		ADVANCED		EXTRA	
BANDS	CW	PHONE	CW	PHONE	CW	PHONE	CW	PHONE	CW	PHONE
80 METERS	3.7 to 3.75	X	3.7 to 3.75	X	3.525 to 3.750 and 3.85 to 4.0	3.85 to 4.0	3.525 to 3.750 and 3.775 to 4.0	3.775 to 4.0	3.5 to 4.0	3.75 to 4.0
40 METERS	7.1 to 7.15	X	7.1 to 7.15	X	7.025 to 7.150 and 7.225 to 7.3	7.225 to 7.3	7.025 to 7.3	7.15 to 7.3	7.0 to 7.3	7.15 to 7.3
20 METERS	X	X	X	X	14.025 to 14.15 and 14.225 to 14.35	14.225 to 14.35	14.025 to 14.15 and 14.175 to 14.35	14.175 to 14.35	14.0 to 14.35	14.15 to 14.35
15 METERS	21.1 to 21.2	X	21.1 to 21.2	X	21.025 to 21.20 and 21.30 to 21.450	21.3 to 21.450	21.025 to 21.20 and 21.225 to 21.45	21.225 to 21.450	21.0 to 21.450	21.2 to 21.45
10 METERS	28.1 to 28.5	28.3 to 28.5	28.1 to 28.5	28.3 to 28.5	28.0 to 29.7	28.3 to 29.7	28.0 to 29.7	28.3 to 29.7	28.0 to 29.7	28.3 to 29.7
6 METERS	X	X	50.0 to 54.0	50.1 to 54.0	50.0 to 54.0	50.1 to 54.0	50.0 to 54.0	50.1 to 54.0	50.0 to 54.0	50.1 to 54.0
2 METERS	X	X	144.0 to 148.0	144.1 to 148.0	144.0 to 148.0	144.1 to 148.0	144.0 to 148.0	144.1 to 148.0	144.0 to 148.0	144.1 to 148.0

THE COMPLETE MORSE CODE
COURSE FOR THE PC

*Generate random characters at ANY speed

*Generate random QSO's-similar to the VEC

  exams - at ANY speed

*Send text from any external data file

*Complete lesson on learning the Morse Code

*Includes 32 page book on Code learning and user's

  manual

*Plus many, many more features

Ameco's Morse Code course for the PC is the most versatile program of its kind. It is user friendly and menu driven, containing over 18 options. It will run on any IBM PC/XT/AT (or 100 percent compatible) at any clock speed, in either monochrome or color.

There are many other features, including quiz sessions for the beginner, as well as the ability to alter letter, character and word spacing to simulate HI/LO spacing. The program also will turn your keyboard into a straight or iambic keyer.

This course is ideal for the beginner, and perfect for the licensed ham who wishes to upgrade! All from AMECO Publishing, the oldest and largest publisher of code training material, for over 38 years.

AMECO'S Morse Code Course for the PC (Cat. No. 107-PC)..$19.95

CODE PRACTICE OSCILLATOR and MONITOR

Model OCM-2

- Large 3″ speaker with volume and tone controls.

- Latest IC circuitry for high quality sound.

- Professional equipment at a very low price.

- Converts to a CW monitor. • Kit form or wired.

Model OCM-2 is a professional, high quality code practice oscillator with an attractive two-color panel. Its 3″ speaker and advanced IC circuitry permit a high quality sound which can be varied in volume and pitch. Speaker or headphones can be used. Once the code has been learned, the OCM-2 can easily be converted into a CW monitor for use with the transmitter. Model OCM-2 is economically priced and is available in kit form or wired. Easy-to-follow instructions make kit-building an interesting, educational project.

Model OCM-2K - in kit form $17.50
Model OCM-2W - factory wired $22.95

AMECO EQUIPMENT DIVISION
AMECO PUBLISHING CORP.
220 East Jericho Turnpike
Mineola, New York 11501
(516) 741-5030